THE WAVE HUB CHRONICLES

Lucy Wells

Coherent VISIONS

The Wave Hub Chronicles
Lucy Wells

First published
2011

Coherent Visions
BCM Visions
London
WC1N 3XX

ISBN
978-0-9569367-9-0

Cover Photograph
Janet McEwan

Typesetting and Layout
Jonathan How

Distribution
Edge of Time Ltd
BCM Edge
London
WC1N 3XX

www.edgeoftime.co.uk

> Oct 2012
> For dear Pence
> Wishing you an
> extremely happy birthday
> year.
> Great to see you!
> Lots of love from
> Lucy x

In the hope that we all learn to dream
And that we wake up in time.

November

The morning was stunning. It had just poured with rain then cleared to shine directly on St. Ives, sun blasting like a spotlight, high majestic clouds contrasting sharply against the iron grey sky, icing on a slate blue sea.

The little dog was so shiny coal dark he made a moving black hole as he skipped about on the golden sand.

Our two dogs looked a little blurry by comparison, rough around the edges, recovering from the mange. There's a neighborhood fox who runs the sand dunes or towans as they're known in Cornwall, maybe the mites came from him or her, transferring from animal to marram grass to animal to burrow into flesh and dislodge fur follicles, a whole world of animal, insect and plant activity as good as invisible to the unaware human.

There we all were strolling on the beach , the humans among us dressed for any number of practical jobs of the day; the chopping wood, the baking bread, the growing vegetables, the sewing cloth, the fire making, the washing up, the washing down, the shed building, the sweeping out (living on sand this is a constant), the chatting on, the driving through, the making up of all the stuff of a simple quiet low tech life.

We start to chat to the owner of the ultra black dog. His name is Patrick, he is an Irishman who tells us stories of his family life in Galway, as the landward horizon colours with reflective fluorescent yellow figures.

"Hey, watcha doing here? Planning to build a motorway on our beach?" Our new friend calls over to the yellow jacketed arrivals.

They come over maybe a little wary until Patrick's chat meets the ears of a fellow countryman, one of the blokes comes from Cork.

They are engineers come to reconnoiter, looking for where to drill a hole through the sand dunes to push a pipe that will house a cable that will run out to sea 10 miles or so where they will then sink a giant socket on the sea bed, the Wave Hub.

"It's a world first!" says the engineer from Cork.

We stand in this impromptu meeting feeling proud that this first thing is happening, that technology is being harnessed for such a wise and clever purpose, making electricity from the beautiful movements of our Grandmother ocean. She need only ripple her waters and a thousand lights may sparkle. One tide could power a myriad washing machines, each wave a million plug sockets.

Computers, televisions, laptops, desk lamps, irons, kettles, sound systems, hairdryers, liquidizers, coffee grinders, ovens, razors, routers, radios, vacuum cleaners, all powered courtesy of her magnificent currents.

I feel thrilled.

The drilling will be going on directly underneath our chalet, from the old power station over the back, through to the beach. The wave power transported through the dune will ultimately surf through the grid.

The Wave Hub Chronicles

The Old power station was a coal fired affair with the grey concrete look of the third Reich. Decommissioned sometime in the early 1960's, the supply of coal was no longer needed. It used to arrive by boat or train. The quayside infrastructure and workers then became redundant. Now all that remains is a long wall that used to prop up the conveyor belt that ran from the ships to the harbour, some iron tracks extremely hazardous to the less than cautious bike rider and a derelict old building with smashed window panes and graffitti.

Next to it sitting within a barricade of fencing is the modern tranformer, power arriving by pylon then underground cable, it carries on beneath the river to the end of the land.

This morning was heartachingly beautiful. So clear, bright and light with the fishing boats gaily bobbing the waves in their sparkling primary colours and puddles gleaming reflections back up towards the brilliant blue sky. The slanting sunlight of November sharpens everything with its particular angle. So different from the straight up glare of the summertime light, the angle creates its own wintry perspective.

In the summer months our beach fills with eager holiday makers. Whenever the sun shines they come and lie down together in the presence of ocean, sky and each other. Little camps are set up with wind breaks to keep out the neighbours and barricade the wind. Microcosmic versions of the cities they come from are reconstructed on the sands, resonating with lost nomadic memories. Here, where it is finally free once you've paid for the petrol, the chalet or caravan, the wet weather contingencies and stocked up at the supermarket, finally there is rest from the cost at the ocean's edge, this precious moment with an elemental being, listening to the waves, jumping in the water, watching children slow down or speed up in the absolute rhythm of nature, zoning out with the sea fresh air, as the tides come in and the tides go out.

Here, where all those bodies lie on rare hot summer days, this is where they'll dig. Beneath the sheer bank last week's stormy high tide pushed up to six foot, dropping straight down into the clambering waves. When I lived inland I would often dream of a sheer steep ocean's edge, waves battling, threatening the dry spaces with wild water.

The pipe will go under all of this, six foot under as a pipe grave it will lay. A man stands there now filming his two young daughters with his mobile phone as they hit buckets with spades in the precious morning sunshine.

This morning clouded over, greyed, darkened, blustered and squalled and brought in lashings of rain, changing the day and the feel of the place completely within moments, eternal and constant, yet continually changing and in flux.

Today it is so still and moist without actually raining and the tide is so far out that I worry that all electrical appliances powered by these conditions would spark, smell of burning plastic and simply give up the ghost.

The extreme stillness opens vast space in the landscape. I move slowly along the estuary seemingly miles to the distant white foam, passing nine swans quietly sitting on the water. I wonder which one of them is without a mate, lonely or bereft. Lonlieness is so endemic with so many people on their own the swan momentarily embodies it and a wave of sadness washes over me.

After returning landward from the waves I see a pool of such mirror stillness reflecting the sky and slanting horizons, the dunes topped green with marram grass and Lelant church tower surrounded by the golf course. This perfect replica is disturbed by the dog and the church tower wiggles. Turning now towards my cave the pool appears suddenly empty, cleared of any message, like a misted over scrying pan and the damp atmosphere materializes into rain. Beauty and sadness swell and flow out of my eyes as salty sea tears. Grief. If we were really to consider all that has been lost in our world we would be overwhelmed by grief and unable to live. I consider the layers of protection, defense and emotional projection necessary to live with the way it is in our world. The faces on the high street, the stories I hear told tell me all I need to know about the state of things.

The storm hit and I understood that this had been the intense sad calm before. It rained sheets and buckets then started to blow and blasted and slammed and shook us about in our bed all night. The wind shrieked and screamed like a demon.

In the slowly declining wind that followed I walked past the old power station where the tranformer sits gridlike behind its metal spiked prison fence. All square and grey with its ceramic mushrooms it hisses and crackles.

Engineers are congregating. Yellow jackets, hardhats and steel containers are appearing over the back on the scrubland around the transformer where evening primrose, vipers bugloss and blackberries abound. They are putting up steel fencing rectangles and walking about on the beach, looking.

Tonight the lights were blazing in the substation building. Three rectangles of orange illuminated the fizzing terminals. The men had gone home, it is gathering dark at a quarter past five. It's still very windy although the rain has stopped. Friends and family have rung to see if we're alright in the storms because they see on the television that we're taking a battering. It's kind and I'm touched but the drama of the TV weather report seems unrelated to the actual drama of our existence.

I met Patrick again with his black dog this morning and he asked me about the Tai Chi I was doing on the beach. We talked of energy, gravity, heart, mind, how to relax and the Chinese concept and philosophy of Qi flowing in the meridians of our bodies, creating harmony and good feeling in our internal organs, emotions and spirits.

If we were to look at our bodies, minds and spirits as composed of five fundamental elements that feed, nourish and control each other in order to create a healthy, balanced harmonious whole, we could also apply this to the world we inhabit.

Fire, Earth, Metal, Water and Wood are the forces that constitute the Chinese 5 element theory and represent every aspect of life and relationship that govern our existence.

Vital life force, that which animates or Chi (Qi) as it is known in Chinese, flows through all nature, including us.

When all is flowing smoothly and without blockage it allows us to experience our existence as in harmony with all nature.

The fact that most of us have had our connection to nature so severely compromised by generations of industrial and financial manipulation, our ancestors having lost their rights to the Earth through successive acts of disempowerment by the greedy side of human nature, double binds us into defending against Spirit, Chi and enormously powerful elemental forces, often ignoring their possible existence , validity only granted to the seen and tangibly proven aspects of this great mystery we inhabit.

However we may choose to relate to our existence it's becoming nigh on impossible to pretend that our human endeavours are working out well as we search for sustainable options.

The tide was far out white lacing the horizon in time with the waves breaking. They haven't started drilling yet, just setting up over the back.

The Wave Hub Chronicles

The wind was howling again last night, my ear is aching perhaps from the sound of it. Its not cold just extremely blowy– blowing enough to have to force your body forwards against the wind, not so much that when you lift one leg up to take a step you're blown backwards but getting near that way. The dogs' fur is blasted back on their faces making their snouts look extra long and pointy as they run. Sand whips into my face so I keep my head down and plough through the dunes until I can turn the corner at the estuary and walk with the wind at my back. The dogs play push me pull you with a stick spurred on by the gusting air. There's spray flying off the white tops of the waves, elegantly fluttering, dainty ostrich feathers on a Victorian lady's hat. Where water lies on sand the wind patterns the liquid wrinkling along its surface. The urgency, insistency and sheer overwhelming energy pushes me to a point of listening to as much as I can bear. The power of the wind is so magnificently inhuman it's hard to hear and every moment with it is an effort.

Away on the landward horizon three fluorescent figures appear, underneath the church topped dunes. They push through the air, separating and looking at the dunes with engineers' eyes. They gather at the bottom of our homeward slope, talking. They are wearing plastic goggles to protect their eyes from the sand, I recognize the one from Cork and another from Liskeard. We greet each other although they look like they'd prefer not to, they're at work after all.

Wild winds have always blown and rain lashed. And scorching sun has parched future deserts and icecaps melted and refrozen after scouring through vast swathes of our world. Everyone talks of climate change. Clever academics write theses, scientists put forward various future scenarios, protestors act directly as hurricanes blaze through vulnerable humanity without so much as a backward glance. Tsunamis gulp down entire coastal populations, floods, droughts, fires, earthquakes, volcanoes, resultant starvation, thirst and disease ravage

The Wave Hub Chronicles

our peoples and still we think we're in charge. There's a
kind of madness at large with our relationships.

There's been a bit of a reprieve in the weather, it's calmed
down with a light breeze, blue skies and fluffy white clouds
but big winds are forecast for the coming days...

I awoke with a start in the night. Through my sleep I'd
become aware of an extraordinary still calm that spread
through the chalet like a ghost. Eery to be so still and quiet
after the racket of the wind.

This morning there was a meeting with the engineers and
the Regional Development Agency over the back by the
steel containers, one of which is a store, the next a toilet,
another a dining room/meeting area with a kitchen bit and
then the office.

They had very kindly and courteously invited all the
neighbours and interested parties to an introduction to the
project meeting.

We all met outside and everyone was very friendly and
excited about what is going on. All first names and
lots of positive chat about the Wave Hub Project. The
photographer from "The Cornishman", our local paper,
took the future Wave Hub manager around to the beach
to snap him in situ, the photo would show him striding
purposefully through the marram grass looking out to
sea. I imagined him smiling, pointing to the place in the
sand where they will be drilling, some bunnies in the
background looking worried.

Another man held up a computer generated image of
the place out to sea where the wave hub will be situated
in its own 4 by 8 km enclosure. This marina will house
the Hub and the power buoys that will be plugged into it.
The hub itself is about the size of a mini car and will lay

down on the sea bed weighed down by rocks. Wave motion will power the buoys. These technological inventions are as yet still in the experimental stages of existence in the private sector. There have been practice runs up off some Scottish shores but never yet any linked up to the grid. 25 km of copper cable will bring the energy back from the Hub to land, through the sand dunes and into a new substation where enough electricity to power an estimated 7500 homes will be received, transformed and fed into the national grid.

They need 25 km of copper cable to reach the marina 10 nautical miles away as they have to follow lines of seabed rock and circumnavigate various wrecks that pepper the coast line around here.

They'll be digging a pit and trench in the beach to receive the cable from its sea faring adventures before it burrows through the dunes and goes on its way to light up our lives.

As I left the group to climb back over the bank I met one of the engineers busy digging a trench and erecting a low barrier.

He gave this response to my quizzical enquiry:

"It's a reptile fence, seriously, to stop any reptiles from entering the zone. English Nature requested it."

I wondered how the lizards and snakes managed before this bit of reptilian health and safety and whether or not they might think again before perilously simply slithering around the edge of the fence which finished a few metres along a bit.

The storm continued all through the weekend, lashing rain and wind until the craziness permeated the chalet walls and delirium and argument reigned.

As massive rains pour, huge areas of the North West have flooded.

The Wave Hub Chronicles

The Wave Hub Chronicles

Just up the coast at Gwithian, a very popular surf break made an article in" The Cornishman" due to despicable images of sewerage flowing into the sea via the red river. When rainfall is high the shit literally overflows. Whoever thought it a grand idea to pump our excrement into the Sea? Thomas Crapper was keen on water closets but when you stop to think about it, it's a mad idea. Why dilute pooh when you can use it to nourish the Earth and it pollutes the water you dilute it with? My friend, on enquiring of the sanitation authorities the situation regarding sewerage, was informed that before a local authority investment of millions of pounds the effluent had been the strength of whiskey it afterwards came out a bit more diluted as lager. Could this be diagnosed as irritable sewerage syndrome?

Today they brought in the massive machinery. Three gargantuan vehicular monsters were pulled up on the quayside waiting to enter the engineers' compound. They carried the extremely large digging and drilling machines to be deployed on the job. As I ventured onto the beach despite the wind, a dumper truck was reversing homeward, yellow light circling atop the cab. They had left another version of the reptile fence in a kind of horse shoe shape, a 50 metre arc at the back edge of the beach before it rises up to dune and sea buckthorn. Just behind it another small barrier had been put in place made out of that orange holy plastic danger men at work stuff. It looked unusual, unnecessarily guarding a line of square stick pegs that pointed up the slope towards the future route of the under sand tunnel.

The drilling started this morning with light tinkling sounds throughout the chalet as plates, cups and glasses jigged in the reverberations ricocheted from the deep underground boring, I could feel my teeth vibrating in their sockets. The yellow jackets looked like clichéd boys with toys on the beach. One sat with a tractor and trailer outside the fence where they are digging a large hole in the sand. Long lengths of iron girder lie in wait to hold the hole together or apart or in place once dug. Other men watch and measure, looking about from under their hard hats. Another sits in a digger with caterpillar tracks digging in the sand on a windy beach.

More and more heavy plant is being delivered over the back as activity hots up and puddles get reinforced by the many comings and going and the track back to the quayside and town behind shines slickly with mud.

Hayle town with its harbour, estuary, tidal pools, many quays and Victorian sluice gates is the site of great Cornish Industrial heritage. The foundry, the Tin and Copper Mining Companies, the fishing, the tourism all have sucked on the nourishing breast of Mother Earth's resources in this rugged, wild and beautiful place one of the largest and warmest river mouths on the North Cornish coast. Without the harbour alive and happening the town seems to suffer from a split focus. There are in effect two town "centres" placed along the ribbon of road that flows from the tidal pool at the Copperhouse end to the train Viaduct at the Foundry end a couple of miles in length. Up behind this road in land far from the long wide beach are rows of houses and estates that have mushroomed over former fields, grey, concrete and close. These house the generations of post industrial Cornish and others that have moved here from upcountry.

Upcountry is definitely a different country, Cornwall being its own land far and away from the main body of Britain, relating more to the Sea and beyond than the inland areas.

As I walked along the river towards the beach in the welcome lack of wind a single swan and a single egret were peacefully gleaming in the morning sunshine. Two white birds in the absence of fury.

I came around the corner into shadow and in view of the digging site, which looked a mess where the boys have been playing.

Hammering metal struts with great big bruising bashers, the dumper truck and tractor and trailer sit by and watch. Tracks across the sand define the patterned roadway to teabreaks and lunchtimes.

Moving further away the pit and its enclosure, otherwise known as the coffer look small and insignificant on the huge wide beach and are as nothing compared to the horizon, the noise an irrelevance in the wide open space.

I arrive at the rocky cave for my Tai Chi practice.

Each weather, each internal circumstance ,each external environment affects the flow of the movement, the breath, the ease with which to allow the simple real conscious awareness that is essential and not affected by these variables. Sometimes energy feels to me like treacle, sometimes froth, sometimes rocky boulders, sometimes greased lightning, always a dance on the sand between rocks, sea, sky, clouds and me.

This morning in no hurry I climb up and over the rocks and pass a man pruning sea buckthorn on the top of the dunes to clear the view for the chalets behind,

"They like to be able to see St.Ives when they're on their holidays."

The same man had planted the sea buckthorn about 25 years ago when a couple of women had brought back a cutting from somewhere on the east coast and it has thrived and done a great job stabilizing the dunes and fruiting orange berries of amazingly high vitamin C potency.

Arriving home I discover we've been quoted in "The Cornishman" as supporters of the Wave Hub.

"I am delighted" says a curious local, "The project promises great things..."

Meanwhile over the back the massive drilling machine with its dyna rod segments is getting welded up in preparation for whatever next. The rain starts to fall anew and the puddles reestablish their territories I wonder where the foxes and reptiles are living now.

The wind is back, swung around a bit to the west and colder. It's the first time all year that I've considered wearing gloves. Three swans languish in the estuary.

I crouched and stretched during my Tai Chi practice, embracing the sky as my shoulders and spine released lumpy resistance.

There is a load of action on the beach this morning. The hole is now surrounded by the tall iron girders sunk in at various depths, which give the construction a castellated air. Two sentries in orange boiler suits with white hard hats stand guard at the entrance. They happen to be exactly the same height so lend some complementary symmetry to the unevenness of the pillars about them.

I chat to a cold guy shivering in the wind. He's waiting for his fry up.

I spot the drill bit poking out from the side of the dune, speared through from over the back.

The orange crane lifts a large black hoop of plastic piping up to the sky and deposits it down again as the orange men attach another big bit to the end of the drill so they can fix it to the end of the black pipe and haul it back through the dune. At the moment there are two drilled through tunnels, one to take back all the slurry from the one that will eventually house the pipe and cable.

The Wave Hub Chronicles

Not possessing a TV we missed the piece on "Spotlight Southwest" yesterday. It featured an archeologist standing by the coffer.

It was hard going on the beach with the wind, rain and sand firing through the air, one of the blokes told me, describing the day.

After a brief interlude of calm it's started blowing up again, squalling as sloshy wet as ever, raining, raining, raining, the beach works look all topsy turvy with sand piling up against the coffer, big drifts blown over the coils of black piping, the wind's blown everything over and under and through and around.

Flung far from the Wirral and London two more engineers arrive on the scene. They will be building the new substation and have come to look at the progress so far, I tell our main man that my teeth shook when the drilling was going on, he seemed defensive in front of his visitors although I assured him I wasn't complaining and not about to sue for dental treatment. I imagine my teeth flicking about in their sockets as the gums perish and a skull prevails, our mighty concerns mere grains of minutest sand in the vastness of existence. The archeologist hadn't seen anything noteworthy in the sand pit; there were no buried temples, broken jars nor marketplace corner stones yet the dunes have been home to humans for thousands of years their bones contributing calcium to the silicea. More recently the landscape has been a popular dump, rusty bed springs and engine parts sticking out all over the place at odd moments.

Living on sand increases the sense of transitory time. Its movement gives it that famous shifting quality, its not necessarily stable or steady and yet it does accommodate, it packs in, it drains well and is comfortable and cushiony temporarily, at least.

There were three swans in the estuary this morning. I wondered which one was the gooseberry. Like a husband, wife and her sister at some garden centre café on the outskirts of town, they bobbed about in the push of the river. Further along a funky little black bird flipped suddenly over to dive and fish, popping up again it innocently ladidahed on the water's surface before disappearing under again like some pickpocket on the high street–flash and its gone.

All day there's been this deep groaning squeaking sort of noise coming from the depths of the dunes something like bass polystyrene friction. The chalet above is acting as a resonator. I think they're shoving a tube through a hole somewhere. The morning was briefly blue and beautiful. Not raining, sunshine lit up distant gulls and spray peeled off the rolling waves. The dunes and cliffs shimmered with rainwater rivulets that puddled into sand trenches then sank away under.

Nine large black tubes of piping lay gathered on the beach by the estuary. They are that incredibly tough plastic stuff that may never decay even after millennia underwater. I tried to play a game with the dogs and a ball but they didn't want to and it seemed a bit cheeky anyway for them to scamper through the virgin pipes.

Last night was full moon and tonight the tide is higher than it's been in a while, seven metres or so and magnificently the ocean has swept right up onto the shore and is engulfing the coffer and fenced off work site. All the fences on the sea side have been washed over and twisted by the waves' claws, the pit is filling with brine lit by the orange glow of the floodlights the rest of the machinery has been moved as high up as possible, cowering from the encroaching tide. I talk with three of the engineers who are there in the darkness seemingly beyond the call of duty, watching the effects of the washing down.

One says it would be embarrassing if any of the plant got damaged. The tide is due to get higher over the next few days before it gets relatively low again and then they're out of here by Christmas...

It all looks very small and vulnerable in the face of the advancing waves. The fluorescent jackets look garish under the floodlights, desperate, like a grin on the face of an acutely lonely person.

They stand utterly defenceless while the effects of their work, the months of planning, the years of financial bidding, the boardroom dreams get licked and splattered by the very central player herself.

This morning's early high tide had swept away the remains of the seaward site fencing and in league with the bright westerly wind had splattered sand against the flood lights, generators and piping. I got to peep into the pit as it was being pumped out, metres deep the vertical iron girders are held in place by horizontal steel props bracing the sides. The guy who'd been waiting for his fry up the other morning was overseeing the pump, he'd told them those safety fences were a waste of time but health and safety protocol often defies natural common sense.

I have a friend who would like to host a festival of danger where people could take full responsibility for themselves over uneven ground and with shoelaces untied.

Attempting to legislate against all life's risky moments is similar in futility to puny metal barricades in the face of a high tide with a wind behind it.

Further up the shore I noticed a few jellyfish washed up. The ones with the purple circular patterns inside the translucent jelly. More and more appeared until it was clear that an entire colony had been brought to beach with the sea's high tide surge.

In defiance a few beautiful survivors gently undulated in a nearby pool, mushrooms in refuge.

My Tai Chi unleashed a thought about the shortness of life. A childhood friend has just had his funeral. How courageous are the ageing, moving along in lives knowing that they are finite. We've got to be brave to truly be alive in the ever changing river.

This morning was still, misty and moisty, the sand like melting snow underfoot. A bright blue ball lay in a puddle, we kicked it around until the wind took it and chased it off to another puddle where another man claimed it for his dogs, he called over to ask if it was ours, by that time I was standing chatting with Patrick and we replied it was a gift from the sea.

We chatted of life and death in the philosophically post catholic mode familiar to those who have been through that particular mill, and decided that God can't just be an external entity but is bound also to come from within and how we need to consider Death's place at the table of Life, until the dogs started barking so we went our separate ways.

I walked towards the coffer passing a washed up log covered in goose barnacles quivering and quietly crackling in a writhing mass. Like aliens on a floating planet they open their goosey beaks to reveal a spiral formation of a tongue extending slowly out then in again in a slow motion gasp.

Not much happening on the works front except the pump is busy baling out the brine after another high tide. I reckon they'll hold off doing anything more until the risk of infiltrating waves is passed.

December

Coming back from the furthest place of the low tide, from a play in the waves in full neoprene armour, I pass the works. Seeing no one on duty I go in to look into the pit again where water still fills the lowest third, perhaps it always will being that low down in sand. I'm cold and wet, the iron grey sky throws rain at me and my body board so I turn to go up the slope. A young man in orange boiler suit emerges from the green onsite container, we talk about the sea, diving and surfing. He's down from Bolton and is dealing with the slurry.

"It's environmental" he says in a cryptic kind of way, as if his relationship to that word is somehow ambivalent.

The goose barnacles are dying on their log, which has been repositioned by a number of tides, none of them strong or high enough to refloat the colony. They are slowing down with their spiral protusions and wriggling earthworm bodies, gasping for seawater as we would for air.

The Wave Hub Chronicles

In the blustery blue afternoon I noticed a number of guys in the garden at the top of the bank, territorially overstepping invisible boundaries. And it is a Sunday.

One is from Pontefract and is laying out a wire to guage where to drill. Another is also from Bolton and is the father of the son on the beach. They know all about drilling. How to do it. Practically. They are the ones who do the actual work, in demand all over the country. One of them is going back up to Cumbria tomorrow to help repair bridge damage caused by the recent floods, I gather that the drilling so far has only been to put in the slurry pipes not the actual cable holding tunnel.

These guys are tough and hardy, they exude experience and carry themselves as experts. They look like they could be from some goblin tribe, underworld burrowers who know the earth in a very particular way.

The very high tides are over for a while so the drilling can continue in earnest. They'll be drilling through with one size bit then gradually increasing the circumference of the hole until it is wide enough to drag the large black plastic pipe through the dune. The sections of pipe, which are laid out by the estuary, all 15 of them, each measuring about 20metres will be welded together to make a great long snake that is then poked through from one side of the dune to the other, a giant piercing in a sandy lobe. We will audibly witness the penetration of our mother Earth as the sounds resonate through the chalet.

It is very wet rain today. A fiddling, infiltrating kind of wind pushes the wet through clothing, skin, flesh to very bone. Wetness hangs heavy in the air, its deep down damp, not at all dry.

The Geordie contingent are on the job welding the segments of black pipe together. The weld is so strong the pipe would disintegrate before the join gave way, so they proudly tell me.

The Wave Hub Chronicles

There's a smart green machine standing on the sand with two sections of pipe in precision position waiting for their union. They are about to erect a tent over it all just like a wedding reception except it's a welding one. Much as it is such a happy occasion the guys working would like their next job to be in sunny Spain they say as the rain pours down their necks cricked against the wet wind. I suggest the heat might produce problems of its own such as dust in your throat or sunstroke. As rain drips off noses its hard to imagine.

The goose barnacles have died on their log. Some were bitten off by passing dogs, left shattered by the side. Blown sand has half buried the windward edge of the log. A dog has left a pile of its excrement splattered all over them. This is what you get my alien friends, when you visit Earth, a slow, lingering death, beaten up and left in a pile of crap. It's only a momentary thought, I can't help feeling sorry that they weren't swept up and taken to float comfortably in life on the ocean's waves.

Through the rain I can't make out what's going on at the site of the pit, a big digger is moving sand about but the space is small and cramped as the tide is high and almost lapping at the wheels of the tractor, trailer and dumper truck.

I move with my tai chi on top of the dunes, the bay spread out before me as a yellow oilskinclad man walks by. He is from Saltash and came originally to drive the bits and pieces needed for the drilling around from the store to the beach by tractor and trailer, then leave. The management soon realized they might need him more frequently given the tide coming in and out with such insistent regularity, so he has stayed on full time.

"Only gets a bit boring during the day" He says a trifle melancholy, the stunning natural backdrop behind him.

The drillers are working their way backwards and forwards through the dune until the hole is wide enough for the pipe, in a pipe cleaning sort of motion only far slower. I can't hear the Earth groaning for the sound of the rain.

They hit rock or something.

It was while we were standing by the drill on the beach chatting to the drillers that the twisting metal rod ceased turning. It was twilight, the flood lights were lit, the scene was a beautifully contrasting set of sharp images I wished I'd brought a camera with me.

Chewing the evening breeze with our friend, the main man from Bolton, we were stood next to a pile of the drilling rods as the boss engineer warned us to be careful of the grease on them getting on to our clothes. He himself wore a collar and tie underneath the regulation fluorescent jacket and would definitely not have welcomed the stuff on his own clothes but I think he was also wanting us to go away as we were riskily close to the centre of operations, the sea had done away with the safety barriers and the operations were his responsibility at the end of the day. We cheerily reassured him that we weren't worried by a little grease and the next rod was inserted as the previous one disappeared into the bowels of the sand dune, with the help of a little digger, a belt and the guidance of the drillers. We didn't want to miss this tiny episode in the total industrial drama of here and now on the beach.

It was then that the drill simply stopped. In an emphatic lack of movement nothing happened. Following this full stop all attention went elsewhere.

It wasn't until the following morning that we heard that they had met with a proper obstacle all those metres deep, unplanned for hard rock or something, not the sandstone that may have been expected, something resistant enough to ruin the highly costly drill bit. The proceedings thus delayed, decisions would need to be taken in the portakabin office while the drilling men waited on the beach.

No one thought to consult local knowledge although rumours immediately abounded regarding all the many things that have been buried in the dunes over the years, including any number of sea going vessels and a large variety of miscellaneous mystery items tipped and dumped during eras where the edge by the sea was not an SSSI and industrialization was raging only yards away inland. Those kind of stories are mere irritants when important multimillion plans are being threatened by unidentifiable buried obstacles or UBOs as they might more commonly be known.

The Head office honchos met, calling in all sorts of managers from afar to discuss how to execute the plan. Meanwhile the actual executors of the plan stood outside on the beach, incredibly disregarded. This mind numbing lack of inclusion must surely have something to do with the amount of grease on a boiler suit.

The welding is almost completed, an awfully long black snake of thick pipe is curling around the base of the dunes waiting to be inserted from beach through to power station. As we chat about the disrespect for the drillers' experiential knowledge and problem solving capacities our friend glances at the end of the pipe.

"That end should be capped," He says and tells us about a job he worked on somewhere in Scotland.

One of those long pipes had been put in, chlorine pumped through and mains water connected up to the village. Four or five weeks later everyone in the village started getting ill. Tests showed something in the water so they opened up the pipe to find the bleached and rotting carcass of a fox who had got in so far, couldn't turn around and died undetected until its putrefaction had passed on the bugs. I looked at the pipe with new eyes imagining the horror of incarceration in its dark roundness.

The drillers also speak of an unpopular new boss; emotions run high as they describe the worst kind of punitive school teacher and caricature themselves into some naughty kind of enemy.

The weather has passed into a calm, blue phase of high pressure and it's a relief. This morning the sun rises, a golden ball in a socket of cloud and throws a glint to distant St.Ives' windows. We're approaching the shortest day and longest night, light is now precious and limited, its fine quality and brightness is a gift today.

I meet Patrick and his coal black dog on the beach,

"Have you heard the latest?"

"That they've hit rock? Yes!" and I tell him about the Bolton crew, the Geordie welders, the unpopular boss and the hierarchical meeting.

"Well I've a bank man as my source," he says and I imagine one of the newly disgraced thieves from the City.

"Oh no," he says, "not one of them boys. This guy is employed by an agency to move all the fencing about and other logistical stuff like that to do with plant care."

And now I imagine someone tending a giant greenhouse.

When we've finally worked out the banks and plants from their respective synonyms I hear that they've been wrangling over who will fund the delay in the drilling and the cost of a drill bit that can cut through even the hardest UBO, notwithstanding that granite might defeat them altogether, whoever ends up paying.

The Wave Hub Chronicles

The Wave Hub Chronicles

Strangely synchronous, my tai chi patch has now become a rocky place. Formerly a sandy cove, at least a couple of foot of sand have been swept away by recent large swells to reveal the rocks beneath. The Earth is baring her teeth and the sand seems suddenly terribly vulnerable and so transitory even though it has taken all those many thousands of years to become itself in its present form.

At the works site the digger is still scooping sand from one place to another as if on an eternal errand. Three hard hatted fluorescent men stand directly under the swing of the scoop. I marvel at how nonchalantly they stand within inches of skull shattering misfortune apparently unaware of the potential lethalness of the machine. They must be used to it.

One of the Bolton Wanderers is stood next to a Cornish colleague, a bright and friendly man.

"I'm just admiring the view which you are part of. These fellas here would like a long weekend. They've had enough of the Cornish hospitality and want to get home to sort out their wives."

What can you say?

A beautiful weekend of weather follows. The site is all fenced up again and the sand neatly raked like a fancy Bn'B's butter pats. Everything left neat and tidy, shipshape.

What a pleasure not to have the constant thrumming of machinery in the front and round the back, a welcome break as the motorway services say.

Rumours are circulating via our man in the local hardware shop that the whole project has run out of money......

A thick, still, quiet density pervades this morning. Not very successfully I attempt to negotiate my way clearly through the Tai chi form keeping allegiance with clarity despite the many and various wanton thoughts rambling.

The sand is beautifully raked perpendicular to the dunes by water, wind and currents undulating with impeccable creative balance between form and chaos, like a tiger's stripes, totally unpredictable yet seemingly regular. I notice the difference between this pattern and that left by the tidy digger driver over the weekend.

The Drilling machine is back on over the back to stop it seizing up so they tell me. No new news of the UBO.

"Blue Alvin rock?" I suggest in total ignorance, somehow thinking simultaneously of Elvis Presley.

I climb the bank to see the drilling machine at work, it's a massive juggernaut size rectangular thing and in fact its not going and the bits and rods are starting to look weirdly rusty.

A big digger scrapes at mud and gravel, depositing the slush onto a mound, seemingly without purpose, nothing else is happening over there apart from rubbish accumulating in a skip and who knows what meetings in the portakabins. It feels depressed, devoid of energy, grubby and despondent, an abandoned game that's stopped being fun.

Over by the estuary seven swans are swimming. Still an odd one's out.

A cold damp clinging start to this day, almost as longly dark as its going to be, just a few days until the Solstice and with the clouds low perhaps the darkness won't lift at all.

Actually later on the opposite becomes clear. Blue and gold prevail in the short light.

They were pumping water out of the coffer, the pipe was hooked high over the fence so that as the pump sucked, the pipe spewed and gushes of murky sandseawater cascaded out in belches, causing the dogs to leap and bark with excitement at this manufactured waterfall.

Nothing's doing on the drilling front. No News. No knowledge of rock. Just another meeting, they won't be finished for Christmas and there'll be no pipe stuffing through the dune this side of New Year.

Red clouded pink mornings come and go with shepherds' warnings proved false and erroneous as blue days fill the short hours of light. It's still. Winter wins its quiet deserved rest.

So still, so quiet, so packed up and gone away a feeling of defeat, failure and uncertainty is left behind.

A bitter cold wind is blowing in from the Northwest. I'm glad they don't have to work in it. It is now the deepest, darkest time.

"Don't poke about inside me while I'm resting "she has told them. "You wait until and if I'm ready."

And they have gone off tails between their legs to farflung corners.

The Wave Hub Chronicles

January

Half the new year's blue moon sits in a magnificent sky. Large pink tinged clouds fluff up sharply in the freezing morning.

Shards of ice lay wounded by the crunch of car tyres through puddles in the car park. No one about. Too cold.

Strong yellow light beams on the swollen river. The high tide is at its highest, only a few metres of sand between the lapping estuary and the long black snaking pipe where sand drifts mount at its curving edge. The "No climbing on the pipes" signs are frosted over and as good as irrelevant.

A line of seagulls sit around the corner, the horizon a steely grey knife edge cuts a line above their heads. As I approach, they fly.

Walking along the tide line of dirty debris, kelp and bladderwrack mingled with plastics, bottles, cans, jar lids, string, syringes, plugs, shoes and oily wood, the lighthouse comes into view. A blob of rainbow winks in the grey. A fishing boat arrives with a squall of seagulls. The contours and shadows of the cliffs minutely contrast in the rising morning sun.

The coffer compound is deserted. Abandoned by humans the high tide waves have washed the rickety fencing and hung up their sea weed next to the strings tied shanty style to keep the whole thing together.

No one has returned from the holidays. The quiet beach has returned.

Dark grey to steel to turquoise slope in from the horizon to the steep sand bank where I dance my Tai chi this morning with ease, my numb and frozen fingers return to warmth as the movement reaches them.

Over the back the drilling machine is disconnected, the slurry pool frozen over. Only the tinkling of metal and an open gate suggest any life. I spy a single hard hatted figure behind a steel container.

A long list of all the schools closed because of the severe weather(snow and ice) was read out on the radio. We don't get the snow here so no chance of frolicking in it.

I battled across the beach in the bitter North wind. Sharp and cutting it sliced through my layers of warmth and forced me to bow to its mighty chill. I sat above the coffer as the high tide waves baffed against the fencing, the icy sea boiling. Sitting in the marram grass out of the full force of the wind I watched.

In human realms you can watch a person walking, talking, doing whatever and you can interpret the interactions, whether accurately is another question but in some way they are decipherable. In Nature it is other. I cannot read the ocean, I have no capacity for something so enormous, all I can receive is the merest twinkling of a nano second of human thought reflected in a sea of mystery. It's the same with the winds and the sky and the earth, I do bow before them, it's the only position I can take with any sanity. The rest seems like madness.

Yesterday at a distance I saw a digger was scooping sand from the edge of the estuary and shoveling it into a dumper truck. This then drove back to the quayside to dump the sand in piles from where it will be taken to who knows where or what. Cement factory? Agricultural fields?

The Wave Hub Chronicles

I feel angry watching this theft and someone is getting away with it. Barefaced, redhanded, scot free. How many places in our world have been so pillaged? Pretty much everywhere for one thing or another with profits making it dangerous. Taken and sold, stoled, stolen, had.

It seems right and proper for us humans to start dropping our heads in respect for Nature, for the power and the mystery of the elemental forces. To stop fighting and abusing them because there is absolutely no doubt but that they will win. And we don't want to lose.

The state of abandonment at the coffer continues. I'm glad for the drillers that they don't have to work in these temperatures. Mind you they're probably up North somewhere drilling manfully in a snow drift...

Proper winter makes you want to get warm, sleep, go inside, store up strength. Keep it ticking over, don't rev up.

☀

Reports of snow from friends around the country make us envious for the white stuff. And then it comes, fluffing out of the sky weightily floating down from the heavens. Interspersed with lashings of sleet there's not the frosty snowy winter wonderlandscape, more a kind of slush puppy on the sand. Still school is shut and the Lord of Misrule timely.

The abandoned works slip further away into memory and distant possibility. It's winter, lay down your tools and stop.

☀

Glass ground. Rock hard sand. Bright sun shines on ice puddles. School's out. Here's a gift of a day for the school goers otherwise segmented into learning slots, here is suddenly available free time, a deep sigh of relief, freedom and non regimented space.

The Wave Hub Chronicles

We walk into the beach site compound through gaps in the fencing and meet a few of the engineers coming down the frosty slope of the dune armed with reels of blue polypropelene rope and cable ties to reinforce the compound's fencing where the sea and nosy neighbours have been busy deconstructing the barriers.

An article in "The Cornishman" has announced a meeting about the Wave Hub's future progress with the interested local community. They cited that the delay to work was due to the weather, no mention of the UBO. It's interesting to notice these half truths and the associated feelings of disappointment that wash over me in relation to any thing political where it seems no one can be up front whilst busy watching their backs.

Our engineer friend from Liskeard says they are still undecided how to proceed and what to do about the UBO. They can't risk any thing going amiss deep inside the dune as the drilling machine is so phenomenally expensive.

☀

Pink light blushes the frozen beach with cold, clear, quiet beauty. The wave hub remains a cryogenic project suspended on ice.

Meanwhile a big orange digger is still digging away at the edge of the estuary as the dumper truck carries off the spoils. The pink light turns it red. I feel angry at the continued theft and the spurious claim that they are somehow clearing the shipping channel. This activity is licensed by the local council. The license is due to expire hence the rush to get as much as possible of this commodity to buy and sell. I feel a surge of support for the Wave hub's potential when faced with this crass exploitative environmental theft. There's been more coverage of off shore wind farms recently and I can't help but cheer on the sophistication of the power buoys situated in the waves and giant sea bed socket rather than the vast turbines rigged on floating platforms, balancing and prey to the wild winds and mountainous waves, a technology vulnerable to be toppled by the same forces it is trying to harness. Why it needs to be a competition is I suppose all down to money again, looking for the safest, most efficient of the sustainable renewables. Meanwhile maybe it would be a good idea to turn off some of the lights when we're not using them.

We meet a local politician out walking the dogs. He's not at all impressed, doesn't think any jobs will be generated by the project, wasn't very interested in electricity, ultimately it's all just a pokey little substation and an office in town.

Easterly winds bring the polar opposite of yesterday's sublimity. It's a bitter, mean, harsh, dark, slapping wet day. It didn't get light for ages this morning, in fact it was never really light at all. The day was a damp grey shroud.

Last night's meeting was of about 48 people, twice as many men as women, asking doleful questions of the Wave Hub's new manager.

The location is The Passmore Edwards' Institute a grand stone building bequeathed by the benevolent philanthropist newspaper owner to the good people of Hayle sometime during its industrial heyday in the late 19th century.

The boardroom, tastefully decorated with pink drapes, alabaster busts and gold framed seascapes centres around a massive oak table.

Most of us perched around the edges behind the privileged seated to listen to the new manager give a run down of things to date, six days into his position.

Carefully omitting to mention any mysterious drilling blockages he reassured us with candour that everything was on track and that this exciting project was moving forward with enthusiasm with everyone who wanted to be, included.

There is a kind of charisma which oils the aura of the successful PR person, a slick charm that allows us to let them, indeed we relax in the presence of their confidence, the feel good factor oozes forth and we choose to believe the information given. Many political characters seem adept at this particular art of persuasion but with our backlog of cynical history and quietly abusive power structures, dishonest authority has put off so many of us now that the veneer is starting to look a little tarnished.

Actually the meeting was dull. No laughter or smiles, all dour faces asking desperate questions that barely hid other resentments over employment, land ownership and the issuing of contracts. It was made clear that this is a test bed, an experiment, a glimpse of future possibilities. Significantly absent was any mention of the ocean herself, this powerful force watched from the oil paintings and was present in the quavering voice of the old fisherman and the red face of the man asking about the harbour.

She, the now silent reason behind it, listening in the background bestowing either blessings or wreckage. The dim grey meeting suddenly seemed so small and insignificant and us dear people so sick and tired and grumbling.

※

Hardly any sign left of the rotting sheep's carcass that was washed up in my little Tai chi cove. It's been there through the frozen snap in different states of dismemberment and decay. The dogs have occasionally nibbled on it, the meat kept from stinking by the refrigerated conditions. Yesterday it was only a ribcage with a sad little scrag of wool attached. There's not many sheep kept around here, I wonder how far this dead one was swept.

The Wave Hub Chronicles

Today the really far out tide showed up bits of the estuary bed I'd never seen before. An oyster catcher raced a gull low above the water. A tall mysterious looking grey heron sat in the shadows of the opposite bank quietly observing. I considered wading the estuary it was so still and shallow, the sand was sinking but not too deep. All was peaceful until the rumble of the digger approaching broke the spell, then it was right behind me getting into position to start digging out more sand. Escaping the monstrous I strode over to the sanctuary of my cove which the high water had earlier swept out, leaving the sand smooth and pristine, the rocks sparkling, the divine char at work. Poking out from behind a boulder like something stuffed out of view into a cupboard whose door wouldn't quite close, I spotted the rib cage and so did the dogs, so I tai chied to the crunching of bones and the horizon simmering in the sudden unseasonal balm as a sharp wave cut clean through the water.

※

The cold slammed down its iron fist during the night and once again we are back in thickest frosty winter with fog freezing in the air and sand crunching underfoot, as frosted crust gives way to weight. Through the murky atmosphere the sun pushes a patch of yellow light into the landscape. Five swans in the estuary this morning, nothing else appears to move, shocked numb by the sudden fall in temperature.

※

Out on the wide beach one of the dogs makes a beeline for some distant canine silhouettes. The body language of their human companion alerts me to a problem.

"After what happened last time I just want to keep my distance," she shouts over the wide stretch of beach.

I recognize her and remember a brief skirmish between the two dogs, not by any stretch of the imagination a snarling dog fight but a brief snap. I hold my dog still, "so you can walk off safely now" and watch as she moves off into the vast expanse wondering that this beach couldn't be big enough for the all of us, dog territorial behaviour weirdly distorting human relations. Feeling disturbed that such a minor incident from perhaps a year ago could leave such an angry, painful memory I wonder about our relationships in general, to our human loved ones, our pets, our homes. It seems that any perceived threat to an emotional attachment can trigger such aggressive reactions in the current fear infested psyche. We wait like coiled springs in a culture of isolation and recrimination, our precious individualism to be defended at all costs.

✹

It's warm, dark and damp as I march over dressed across the car park. The dogs bark at a figure emerging from the gloom and I greet a young woman. She looks destitute, alone and lost among the dunes. Walking on I start to wonder should I check if she is alright , I start to doubt myself am I uncaring, choosing not to see this desperation or is she simply out for a walk as I am, taking space...?

I do my Tai chi in the near drizzle whilst a raven plays tease with the dogs hopping just close enough to attract them then flapping off tantalizingly close to their advances.

Strolling back past the wave hub site I am struck how quickly an untended thing becomes invisible. When no attention is paid or acknowledgment given, a place or being starts to fade into insignificance, becoming part of the background, an unseen aspect of the world, taken for granted, assumed, downgraded in worth to near oblivion.

Very close to the compound three rocks are embedded in a low mound of sand, positioned with care and precision, symbolic marks made with decision and meaning. The impact of this piece of art smacks these thoughts away as the whole visual image of stones and metal construction crashes forwards. Remember the young woman on the dunes.

※

The mist was thick and it was hard to see far as I ventured out. Hearing became heightened and the St.Ives train sounded like it was very close when I couldn't see to know it was over the estuary, the river as a visual boundary creates distance. With no horizon the near becomes expanded. The fallumph of ball and dog into water rippled loudly through everything.

In this close up vision with no distance to distract my eye, It's shocking to see the changes to the beach. The sand is taken and the rocks are huge. Impossible to track a process minutely but turning around and looking again after a lapse, I see a rocky beach where once there was a sandy one. I know the sea can do amazing deliveries and extractions but this beach has been known for years for its miles of golden sands. As I pick my way across the rocks I feel a loss.

※

The boss engineer was taking photos of the site and stopped for a bit of a chat. He's been working out what to do but now that the sand's gone out of the compound it's going to increase the complexity of the job somewhat. He reckons they'll drill a smaller hole. He kept his face straight apart from the merest flicker of a smile when I mentioned how the new wave hub manager had predicted the drilling would be finished by the end of January.

"We won't be too much longer than that", he offers thoughtfully.

The mention of a smaller hole sends my mind glancing to the big black snake of a pipe still lain curling round the base of the dunes patiently disappearing under the sand.

The Wave Hub Chronicles

※

The fluorescent jacket brigade is starting to multiply. Today there are two striding about, in and out of the compound.

※

An inexplicably edgy day emerges from this sunny morning. Stress fills people and dogs in a calm blue background. A large fishing vessel chunters down the estuary, dogs chasing into the water after a ball, a pleasant chat with an elderly woman and her greyhound but a strange overtone resonates throughout, a tinderbox waiting to spark as the Haitian earthquake repercusses, a butterfly flapping its wings.

※

The transition towns movement that is looking at sustainability beyond the peak oil situation provides all sorts of ideas for living in a responsible way in relation to our environment. Finding ourselves in this strange and stupid predicament where we've come so far from Nature that we are now deeply terrified of her power, it often seems that mainstream powers would prefer to do an ostrich rather than honestly deal with change, investing inordinate amounts of energy into keeping things running as ever, meanwhile there are fantastically committed people everywhere bringing useful technologies and approaches into play. It may be less to do with heads in sand but rather the get it while you can mentality that perpetuates so much of the greedy ignorant business of the world; they after all are only mirroring our own inabilities to adapt and our own greed and complacency.

All sorts of conspiracy theories circulate, as war and business concerns conduct themselves with minimal attempt to disguise their overt and brutal bids for power, money or control over resources.

The Wave Hub Chronicles

The Wave Hub Chronicles

The in the moment beating of our heart reminds us of the rhythm of our natural clock and the natural order.

If we do not hear the heart as the wise and knowing pulse of our lives we are prey to the tyrant fear that can run amok causing this sad separation from the true power, majesty and beauty of the world. This is what seems so truly savage.

This grey morning I look at the dilapidated fencing around the derelict metal pit and feel nothing. There is no meaning there. It's not hopeful for anything. It's empty.

☼

Still and gently lapping the water is high around the beach. I feel a kind of dullness, a difficulty connecting with myself, a cut off kind of feeling, depressed deadened nothing, like there's a huge gap between me and the world.

This feeling often strikes in the night when it can all get too much, it's tricky to hold out, to keep going, to keep believing that my little life is a valid expression within the whole.

Like a weather front these emotions arise and move through me sometimes getting stuck for a while, becalmed in the doldrums.

Today the stillness isn't crystal clear, it's dull and murky.

Tomorrow another influence configures it all anew. It's tiring, winter, a long time in the dark, hibernating.

My neighbour's lights were flickering in her chalet,

"Oh it's been like that since they put the wave hub in," she says "It gets worse when they do the drilling."

I told her I thought she might have a loose connection, should get it checked out and marveled at how easy it is to pin blame onto anything nearby that doesn't know to defend itself.

Something funny is going on today, there's a cat and mouse game at play. I just caught a glimpse of a dumper truck rounding the corner this morning and on my way back this evening a digger disappeared behind a sand dune just as I came along with the exact same visual timing, as if the day hadn't existed between those two moments , rewind, let's have that again. As the afternoon sun dipped down leaving a cold beach I was left standing by the compound looking at the sand, mauled by the digger's scoop to no apparent avail. A pile of sand lay scooped up from the beach to beside the fence. Someone had been playing. A blue and white flag hung limply in the wind, its writing indecipherable on that account, hoisted onto a scaff pole strapped to the metal girders, as if to say:

"Ahoy! We're back, this is our castle!"

Over the back by the electric terminals two bodies were scraping away with gravelly shovels around the drilling machine, its weeks of laying idle in weighty slumber have embedded it into the gritty soil and it needs a good old raking up of the embers before it can spark back into life.

The long black snake of pipe is now officially redundant according to my new mine of information; an extremely energetic 85 year old woman. I know her age because she told me proudly.

The women I know of this generation, born in the 1920's are a particularly fine vintage. Brought up through the depression of the 1930's, young adults during the 2nd world war, living on through marriage, children, divorce, bereavement, grandchildren and great, still with an active interest in learning, community, involvement and participation. Grand women and mothers, living life in a full and rich way. I am grateful to have learnt lessons at their laps. They have listened and offered guidance in the self effacing way they were taught and conditioned to. Not born into an era where their own authority was recognized, they have picked up respect along the way.

So I reckon the pipe must be history.

※

Freezing cold grey has fallen back down to earth again. Leaden and heavy the cold makes it hard to move fast or easy. I feel seized up and slowed down.

Flourescent men are crawling about on the dunes, hammering in pegs, marking trajectories, unloading another load of black pipes of smaller circumference.

They are starting again, ruling out the negative possibilities, hoping for the best, guessing a game with loads of money, untold waste and a contract to fulfil.

A new team of drillers wander about in our garden. Also from Bolton and the same firm of drillers, this lot say they've come to do the job right and will pop in for dinner if I call them when it's ready. Different faces, same humour. Expert style.

※

Today I've been thinking that there's something not very sensible going on with the scale of this project, it's ultimately the same as all the other utilities and services in our western world, out of our direct control so we are continually rendered slaves to the money economy. On this multi million pound scale a single piece of rock or metal flaws the whole operation. Weeks more wages, machinery costs, meeting expenses, someone somewhere mining copper for someone to get extremely rich from resources that are owned by whom? And how?

Meanwhile just around the corner at the entrance to the harbour there are sluice gates ready made that could power turbines with each turning of the tide. The scale is too small it wouldn't even do a bedside lamp in the whole household's needs if our nation was a three bedroomed semi. But as a local community our needs could be someway met. Initiatives putting the onus back into the ordinary person's realm would not be truly desirable perhaps because it would threaten profits and control and be simply too empowering. Even the latest goldrush of selling electricity back into the grid if you have some domestic way of harnessing it saddles people up to longterm fixed rate contracts with power companies.

In the grey light of this cold day it all looks like more of the same old emperor's new clothes.

Perhaps I was a bit over familiar when I greeted one of the engineers by his first name yesterday as he walked along with his boss, because this morning he greeted me with an extremely formal "Good Morning Madam."

He was with a new character I hadn't seen before and I was carrying 3 crushed cans of Carlsberg and a piece of broken glass so for all they knew I might have been having an early morning bevy on the rocks, got so pissed I broke my glass and needed keeping at arm's length. How far from the model teetotal citizen I aspire to be, quietly picking litter in my spare time.

There was lots of machinery bustle with diggers, tractors and trailers coming and going and everyone busy and industrious as they make their second assault on the sand dune.

I register my own personal loss of interest in the repeat procedure, waiting for the next bit that's new, like when you go out and forget something and have to retrace your steps, the journey's not really on until you reach where you left off.

They were digging away at the black pipe by the sand dunes earlier. The tide was so extremely low I was right down in the estuary with the bank towering above me. The swan family gently scudded by, bird clouds in a river sky and a gang of ducks sat basking in the stillness until the dogs came barking and the irritated water rippled at the disturbance of the peace. I didn't want to see any more diggers scratching away at banks of sand, they seem to have disturbed enough.

Last night we were at a meeting of friends of Hayle Harbour. A mixture of interested local people and councilors discussing the harbour ownership and plans for the future. There is an application to put a giant supermarket on one of the quays and concrete in the former sluice gates that have rotted with neglect during the reigns of various entrepreneurs all hoping for a fast buck and sold out once their dreams became too expensive or otherwise unworkable. How many supermarkets can a small town sustain when there are already three? It seems like nothing short of lunacy, even given the tourist trade, as if another will magically conjure its own mirage of wealth. We are dreaming a nightmare, living a fantastic lie.

The river gurgles by reassuringly breaking this flow of thought as she travels out to sea. No amount of concrete, shopping or bombing the moon can conquer her. She can always adapt, flow or dry up, evaporating into the heavens. This isn't her loss, it is ours.

※

In the grip of an icy wind I ploughed my way around by the estuary deep under a bank of sand etched with current memories. I peeked up from my riverside trench, eyes level with the dusty blown sand mist that blurred the vision of the ground. A little blue and white pvc marquee flapped in the unforgiving air where they are now welding the smaller pipes together. The weeks of meetings have concluded that this is the best way forward, two small pipes to each house a cable rather than the larger one that would have had them both laying side by side. A fifty-ish grand decision. (£42,000 for the pipe itself and £15,000 to transport and weld it.)

During my Tai chi I noticed a man walking forwards and backwards in and out like the tide. Standing for a moment in the huge beach space dressed for the wind, he could have been a toddler waddling around with interest known only to himself, then he started dragging his foot across the sand drawing a line out to sea and the image was complete. He stood and took a proud flash photo of his work.

Meanwhile at the coffer his mate was dumping piles of sand building a little hill. I wondered why.

"It's just the sand's been moved by the sea, the beach has changed and that big puddle's gone and this is so the machinery can see into the hole."

I couldn't help but wonder how long that heap would survive into the future of incoming tides. The delay caused by the bits of drilling equipment being sent from Belgium to Glasgow instead of Cornwall could create any amount of work piling up sand for the sea to wash away each twelve hours.

Sure enough later with help from a North Westerly the waves slapped away the men's work and King Canute was once again drenched.

The drillers are still waiting for their bits, there's no rumbling in the nether regions yet.

It's so windy I have to walk backwards to avoid a whipping from the high speed sand. I meet an extremely nimble one legged man walking his dog in this blasting gale. We chat as we move towards the men welding the black pipes. I tell him what I know about the wave hub project. The welders have abandoned their tent the wind is too fierce. I feel for them having to work in this weather, it's enough to send you crazy being in it for a short stretch never mind working a whole day, that is truly grueling.

There is that point with the wind as every sailor knows where it becomes dangerous.

※

The full moon was setting over St.Ives this morning having been out with Mars last night. There's a light cold breeze, the sky is blue and fluffy and there is total wreckage on the beach, the tide has run amok, the celestial beings must be hungover.

The Wave Hub Chronicles

February

This high clouded morning is almost spring like. Although chilly it's nowhere near as wintry as the places we've just been through on a trip to West Wales. I am taking a photo of the three black pipes, a mother and her twins and as I follow the long snakes around the corner I am literally shocked by the change on the beach around the coffer. So much sand has gone, the edge of the dunes is bitten into and dramatically eroded. There is no longer a path to walk up behind the coffer where once was a ridiculous reptile fence.

The sea has smashed the fencing and this time they will not bother to replace it. The drill bit is poking out way beyond the coffer compensating for this incredibly sudden erosion. This must render the coffer as good as useless, another fifty grand's worth of misfortune.

Expensive learning when you add in the black piping, but a relative drop in the ocean in the world of energy suppliers.

The main drilling man says they should be pulling the pipe through either this afternoon or tomorrow.

"Should be interesting," I say.

"Not really" he says.

"Well you know what I mean" I say.

"Yes," he says, then, "this is supposed to be the easy bit, drilling through the dune, they've got to get 10 mile out to sea yet, 50 grand!" he says nodding his head towards the coffer and raising his eyes to heaven as if to say what a bleeding waste.

The main engineer is lurking nearby, not sure whether to be pleased they've nearly finished or a bit embarrassed by the costly blunders.

His position seems isolated and compromised, the clever achiever who has always played by the rules even when his bosses tell him to piss in the wind, whilst the drillers have more of their own autonomy albeit at the price of status.

The extraordinarily rapid departure of the sand leaves me walking about the beach gazing at the spaces left, the sharp rocky substitutes.

I should say the depth of the beach has dropped by about two metres all over since they started this work here.

The synchronicity is interesting; perhaps it's nothing at all to do with it, simply a large shifting pattern with our human influence an irrelevance– or perhaps it has everything to do with it. Very simply speaking it's all going on in this little measure of coastline, chartworthy change in a 500 metre stretch of sand and rock.

So much of Science has taught us to be deeply unimaginative when it comes to our relationship with the natural world, often to the point that ludicrously we will not notice the signs under our very noses.

However originated, these phenomena have already made a massive impact on the project, and rubbed a little salt in the financial wound...that is after all the place where us modern folk take most notice...

☀

To perfect the synchronous, the beach surveyor turned up today. He comes by every six months to measure the levels of sand on all the beaches from Sennen Cove to Hartland Point up the North coast of Cornwall and Devon. Just when we were wondering where all the sand had gone there he was on the dunes with his measuring equipment. His essential message was that the sand cell is a moveable system, so if it disappears in one place it will appear in another, which is pretty obvious. He didn't really have an opinion about the removal of the sand from the system because I suppose it was nothing more than hearsay to him. Funny that. The local group, Save our Sands also gets very tired out attempting to get so called authorities to acknowledge that extracting large amounts of such a natural resource could have any implications. If a digger were to make its way onto a beach in St.Ives and start digging there'd be hell to pay, however in a post industrial zone without the romantic and artistic associations of its over the bay cousin, Hayle is left vulnerable to pillage but ironically the effects could be felt anywhere in the bay.

I was just having a quick wee in the garden when one of the drillers walked past not ten foot away. We did that thing of non acknowledgement so as to avoid the potential embarrassment of the moment. We met again later, he with a mate and we all greeted each other as normal but I bet they had a laugh at tea break…Hopefully they won't stumble upon me in the compost toilet, getting caught in the middle of a pooh is altogether more serious. Now why exactly is that?

Today is mild and still. Wading towards the sea from the estuary the water seemed to curl and roll like silky hair on sandy shoulders, unusually solid and tangible. The great day for hauling through is today, I was told almost cautiously. It seems this job has become unpopular with the workers, a lot of things are going wrong. It's all so unpredictable, there's too much instability in a pile of sand. The reptile fence is dangling forlornly over the edge of the ravaged sand bank, like a discarded snakes' skin.

One of the engineers is wielding an angle grinder,
"We're taking down this eyesore, this embarrassment," referring to the metal coffer, "it's totally redundant"
Again this whiff of disappointment, the lack of success, ashamed how these obstacles have overwhelmed them.
"Then we won't be disturbing the peace and quiet..."
A local country man, I fancy he felt their activities to be somehow at odds with the nature of the place.

I now have a beautifully flat tai chi spot thanks to some sand rearrangement. My form feels like a dance on stage rather than a strategic manoeuvre between stones.

※

Togged up in full wet weather gear for this extremely wet afternoon we met with the sight of one of the smaller but still extremely long pipes lined up on the low tide beach, pointing far away to sea at one end and into the drilled hole at the other. The pipe was being slowly sucked into the dune at about the same speed as the London Eye goes around, perceptible but slow movement. One of the drillers was stood atop the pipe providing perspective.

"Shame we didn't do this in high summer," said another, "We'd have made a fortune charging for the ride."

The pipe then stopped as over the dune on the other side they took off a length of drilling rod, resuming the slow pull moments later. It had been quite a job pumping the slurry mixture to keep the sand from collapsing as the rods bored, so getting this pipe in position will be a relief for them.

We walked further along the soggy beach and found some places where migrant sand was banked up but further along, where the unfortunate landrover had been caught out by the tide all those years ago, the chassis was more visible than its been for years.

Up until just recently it had only been a small angle of tyre jutting out in the smooth expanse of sandy beach.

The Wave Hub Chronicles

Some years ago a farmer and his daughter were driving on the beach at low tide. They stopped to admire the view, some water splashed onto the electrics, the motor cut out and refused to restart.

Then the tide came in.

At the following low tide the landrover was revealed upside down, embedded up to its axle in sand.

※

It was raining heavily for hours now it's blowing out dry, a pale eternal grey as if it could never be anything else. The winter feels very long now, its been dark for ages. We've just passed Imbolc, the return of the light after which the balance is tipped in light's favour but it still feels like a long haul until some easy warmth and late light.

I've walked back past the cave in the black rocks where often there is a man standing with a small fire and a comforting bottle. He seems to be a guardian of the place, back to the rocks, facing fire and sea, at home by his hearth.

There is now no sign of the black pipe that was sucked and swallowed into the belly of the dune yesterday.

Simply sand.

Everything is neatly buried. Soon I suppose they'll cut down the remaining metal coffer walls. Behind them, where the dune has been carved away I spy the rogue remnants of fencing posts. Apparently twenty or so years ago they were worried about the dunes eroding and put up fencing to stop people climbing on them. Some very short while later the sea bought in a huge amount of sand and promptly buried the whole lot in one grand generous gesture.

※

The half waning moon hangs proudly in this morning's clear blue sky. After a night's full on rain the air is clear and fresh, relieved of its burden of greyness. Still and bright, a sense of optimism shines with the sun at high tide. The beach is littered with plastic junk and grimy seaweed vomited up in plaster coloured spume. Industrial sized woven bags, the type used for bujlding sand or agricultural fertilizer lie half buried behind the remains of the coffer.

I move through my Tai chi form on the new flat sand stage, edged in by the high tide lapping at but not quite catching my feet. Attempting to release the build up of winter's damp atmosphere I internally stretch to relax and shift a creaky, rusted up feeling. Standing still I watch the approaching waves slap down their foam onto wet sand. Globs of bubbles polka dot the background then blink out of existence as the sand lightens and the water drains away.

Later, in the sunshine five or six men are standing looking in to a newly dug hole smoking cigarettes. As a gang they are unfriendly.

"We're waiting for the drill to come through, should be through in about half an hour. Bit like waiting for a bus."

It seemed they couldn't bear to wait, sitting in the sun on the beach had got too much.

☼

A bitter north wind is blowing again, sharp and cutting, a lazy wind doesn't bother going around you it just goes straight through. A gun metal grey sky occasionally cracks blue. The sun shines clear on all the glory of St.Ives and Godrevy lighthouse where the rocks protrude and the sea sprays up continuous white flumes, roman candles of ocean spurting, white water highlighting the scene.

At low tide I've started visiting a bit of the beach a little further away from the men and the digging for my Tai chi practice.

They are drilling for the second hole and the digger is busy scratching away for some reason or another. It

all seems boring somehow literally and metaphorically,
I guess its the last stretch for this team, they want the job
finished so they can go, its cold and they want to go home.

※

The wind is still blowing straight and sharp from the
frozen North. This morning the sun rises and shines with
a daring brilliance. The water butt may be frozen over but
the blazing being in the sky still comes on strong. Shadows
are thrown into strong relief; the estuary mud is blindingly
gleaming as baby ducks and peewits chunter about.
I follow the estuary out for low tide miles.

A dumper truck passes by, the yellow flashing light on
top touching distant outlines of faraway clouds. Today
the water seems more mysterious than ever, when I really
consider what it is and how it does its flowing wet dance so
continuously, its wonders never cease. Today its as if there
were something hidden in the waves' lapping approach.
I think of the fisherman I saw earlier with his silver 4x4
pickup loaded with lobsterpots gazing out to sea from the
car park, looking for fishermens' signs, whether its worth
the cost of the diesel to make the trip out. He looked at
me with curious sea wise eyes as I held my dog's collar to
restrain her from chasing his tyres.

On the beach the dumper truck is joined by the big crane
digger and a tractor and trailer, bringing and lifting the
drilling rods for today's poking and pulling escapades.
Then they'll be pulling through the 2nd pipe which is still
lying with its mother snuggly curled around the dunes.
After which it will only be the big mother left lying there
redundant, half buried in the sand. Isn't that just the way?
They'll probably chop her up and sell her for scrap...

Yesterday evening the big drilling machine appeared to
have gone from over the back and the crane was lifting
up the portacabin, drilling centre of operations, as if to
look for something underneath. Funny to see it dangling

there, lights still on inside, workers just popped out for a moment whiles it floats around in the sky for a bit.

※

Another bright and freezing morning in the beautiful cold and I meet Patrick and his coal black dog, we fall into step and immediately start to talk about the wave hub's progress.

"I've so little else going on in my life at the moment that when I see a couple of blokes in reflective jackets with a spade, I immediately have to go and find out what they're doing."

I agreed as we followed the big drag mark left by the pipe in the sand, turning the corner we came upon a major operation in the sharp morning sunlight in the wide space before us.

We started to sing the blue Danube waltz as cranes, trucks and diggers choreographed their movements on this massive stage, one of the drillers conducting from the mouth of the hole.

Action, Music, Lights, the visual treat was spectacular. We stood together and proclaimed ourselves truelife historians and what was it?

"I can't seem to remember any words these days, they're all lost to me."

"Archivists?" I suggested,

"Oh yes, archivists" he said proudly, "We were here the first day and here we are now as they complete this phase."

We parted company happy with our roles as witnesses to this great engineering show stopper.

I went to do my Tai chi by the cave with the pipe lying through a pool just in front, its last moments in the sunshine before being buried forever.

Throughout the day I watch the progress, the boss reckons they'll have their bit of the job done and dusted in another couple of weeks, I don't take too much notice as times and dates don't seem relevant in this landscape, but as the final segment of pipe is slurped into the dune assisted by the lubricating gunge the end is literally in sight.

It's freezing cold and the men have been out there for hours. I offer them a piece of chocolate I've found in the pocket of my coat. I hope it's not stale I hardly ever wear this old coat, it's all ripped up from 15 years of wear and seen a lot of gardening seasons, bonfires and clifftops. I haven't been able to throw it away and now it's proving its magic by providing the hearty chocolate, we immediately start chatting.

"We've just been ratting," I say conversationally, in fact it's the reason for the wearing of the coat. I point at my dog, "She's a fantastic ratter, they've been under the chalet and she's seen them off with her persistent worrying, now we're helping the neighbours with their rats.

"Why don't you just poison them?" Says the tall driller, "I live in North Wales and there the council will come and do it for you"

"Where d'you live?" I asked the other guy, getting off the rat subject.

"I'm in Rochdale, when I'm there. We get about a bit, seen pretty much everywhere, amazing how different places and people are, they're friendly down here."

We compare geographical friendlinesses and I walk away wishing them luck.

As the sun sets on this sparkling day I pop up the bank to look over the back and see the vehicles parking up, slurry tankers now at rest. The two black pipes are recognizable poking out of the ground just beside the drilling machine, smeared with slurry.

※

What a beauty of a day, a wonder, a miracle of all the ten thousand things convening with such quality.

The Wave Hub Chronicles

The sun rising strong over a clean and sparkling frost.

The water, bedecked with swans and peewits, shining like a jewel in the sand.

The Sea lapping the air fresh, cold and clear.

The massive beach without another footprint.

I feel so grateful for the privilege of simply being here.

My worries of the moment are merely peripheral in the face of this grandeur.

I am surprised to see a man sitting in the cab of a digger by the coffer. We start to chat, he's only been a week on the job, I recognize a south eastern accent and we swop stories of how we got here, he's here to take the whalings out of the coffer so it can be dismantled. The pipe doesn't look deep enough to me, it's sure to be the target of a thousand buckets and spades.

We talk about the shifting sands and a neighbour approaches, then all of a sudden we're onto theories of power and money.

The man in the cab says simply "It's slavery, the capitalist nightmare."

Like a pebble dropping through a deep well to its resounding plop, an inevitable gravitational understanding occurs in my mind.

Here is the craving for freedom, the resistance to the treadmill.

Friendliness swells in the atmosphere and I walk away glad for this exchange, calling back a joke to the man in the cab.

☼

Returning from a week away I find it all gone. The metal castle, the big black pipe, the machines and all the other stuff over the back. Gone. Even the swimming pool thing is emptied and flattened, a mess of puddles and mud.

The two pipes are sticking out of the earth like old tent pegs.

I walked on the beach marveling at how easily it wasn't there. It's still bitterly cold so I was on the look out for driftwood and tried to dislodge a bit in the sand noting all the surveyors' marking sticks up the bank, thinking I'll have them for the fire.

※

On my way along the beach today I see a digger digging a hole right by the end of the pipe. A blue boilersuited, hardhatted man is marching about for some purpose.

The disinterested digger driver from down the quay sits fat and blank in the cab. A couple's dog jumps into the hole. They shout.

"It's a great big hole," I say,

They look at me with dislike, shocked that I've spoken out loud, something so stupidly obvious. I must admit I wonder myself.

I notice the piece of wood I tried to dislodge yesterday was marking the hole.

The new men here are strangers on the job, they don't know who I am, nor that they are as good as in my front garden. I feel affronted and decide not to talk to any of them any more, ever again, if this is how it turns out after months of careful watching.

※

This morning the wind has swung round from the East. It's grey and raining in that edgy way that makes every thing feel a bit fast and stressed. The river is flowing fast, flocks of ducks are flying in gangs over its racing waters. The tide encroaches forwards, long dribbles insinuating themselves onto the dry sand. All at a pace, pushy and restless.

Nothing but digger tracks and freshly turned sand remain from yesterday's hole.

Over the back the huge digger scrapes away forlornly at the deserted enclosure, a lonely thing, scuffing away in the corner of an empty playground. The circus has left town– the green sand sifter, the drilling machine, its kiosk and racks of bits, the portacabins, skips and vehicles all up and away to other jobs in other towns.

Shockingly I feel consumed by feelings of abandonment, anger, grief, humiliation and sometime later a sense of resignation and learning, of gathering my will for the next steps forward. Like a theatrical or therapeutic role play these emotions become overwhelmingly real, triggered by some feeling looking to replicate itself.

I feel that huge desolate space which follows being left and shows time as carnivorous oblivion waiting to swallow me whole, misery and all.

I sit with the rocks, grateful for their calm abiding. Between the rock and the hard place is the rock's incredible dependability. Being rocky is essentially so stable until life dashes you against them of course...

The grey, windswept beach encloses around me containing my darkness, wrapping it in its own until we are one being gently rocked by the waves.

☼

It is a grey raining morning. The dogs are barking like maniacs as I spot the orange digger scraping away at the sands. This scraping seems irritatingly interminable. What are they doing now?

A voice from the cab tells the dogs not to walk on the newly smoothed sand, I'm ready for battle then recognize fondly my favourite engineer.

We chat about the clean up job, this is their final handover day.

A bevy of fluorescent men walk over the rise of sand, who's this then?

"Here come the children." Says my friend who looks to be nearing retirement.

The Boss shakes my hand and introduces me to the guys who'll be building the substation and dealing with the cable when it comes in to land.

I leave them to their business as the dogs are now barking so madly I wonder what they're trying to tell me with their noise.

I thought of my engineer friend and how it must be for him working for bosses who could be his sons and how he carried himself with such quiet respectful assurance and was the one who saw and respected the natural beauty of the place and knew the folly of pushing against the elements. How often it is that we don't have the appropriate people in charge. How little regard we have for genuine age and maturity. How tragic this is for a culture.

So we said goodbye and really I should have liked to give him a big hug. I knew his wife had been in hospital and that his back had been hurting and we'd been comradely these past four months albeit distant and beset with the myriad unspoken human etiquettes that figure in relationship. Besides, he was in the cab of the digger.

I just popped over the back to ask if I could have the marker sticks for firewood and they shrugged and said, "Wait til we've gone and then we won't know anything about it."

And that was our final goodbye.

March

The new team are coming on Monday until then I've been reinhabiting the zone over the back that is now barren scraped about mud and gravel with two big wooden block stakes marking where the pipes begin or end depending which way you look at it.

Walking about by the electricity terminals as they crackle and hiss in the rain I think of Tesla, the Serbian electrical engineer whose experimental genius touched directly on the mystical as he attempted to discover a way to directly harness the electro magnetic energy of the universe to power our burgeoning electrified way of life. He was apparently extremely aware of the effects of electrification. Born into the late 19th century along with Edison and Marconi he was responsible for so many of the electrical facilities we now take absolutely for granted. Labelled eccentric and the archetypal Mad Scientist his ideas were indeed exceedingly threatening to the popular capitalist ideals, so no doubt it suited the vested interests of the time to pilliary his inventions as crazed theories rather than having us all receiving the heavenly chi via our plug sockets, live and direct to remind us of our place in the great scheme of things. Whatever Tesla was grappling with in his inner world and attempting to realize with his various experiments the plans for a "Death Ray or Peace Ray" weapon that he was working on, something he thought would end all wars, are still held as Top secret by the American Intelligence services.

So I muse as I collect the blocks of wood from up and down the bank, that will keep our stove going for a while, help see the winter out.

It is blowing from the west today, fresh, high and blustery and the sun has ploughed through the clouds to help it all glisten.

I was alerted to the presence of a dead seal on the rocks this morning as my dog went in search of putrefaction perfume for her neck. A big old dog seal, black and grey lay with its mouth in a groan, eyes already gone and other beaky peck marks along its shanks. I was caught with concern and the wish to let his family know. They breed up beyond the lighthouse. I had just used a photo of them to make a card for a former neighbour who is 100 years old tomorrow and had been thinking of Duncan's selkie stories of the Seal People. Duncan Williamson was a Scottish traveler and story teller who collected loads of stories from the west coast of Scotland regarding relations between the human people and the seals, he himself was authorized to perform Selkie wedding ceremonies... clearly to me it didn't feel right nor proper to simply walk on by when I came upon the corpse.

Since the full moon the tides have been really high. This morning's tide brought up a load of wood onto the beach and rebeached the seal. He had left his rocky grave when I went back to look for him and been shifted further around on the sand where he lay now tagged and therefore officially noticed. We put tobacco in his eye sockets and nose cavity as an offering for his spirit and I went to the cove near the lighthouse and called out to his relatives that we have found his dead body on our beach.

Further along that beach, behind where the coffer had been, the tide has eaten away at the dunes and left the slurry pipe exposed, hanging out of the bank dangling like a floppy lost thing despite all the fastidious attempts by the engineers to control the nature around their interventions.

We've cleared out all the available firewood from over the back and are now watching as the next lot of contractors arrive slowly but surely. Another set of diggers, another day.

The building of the substation.

Royal blue containers have been delivered over the back with red skips and rubbish bins to complement the colour scheme.

They have rebuilt the reptile fence. With all new materials. This level of absurdity is deeply entertaining. The RDA money funds one lot of contractors to put up and take down one dubiously useful fence. The next contractors come and do exactly the same thing. I suppose they could have used the old wood except that we took it and burnt it. So three cheers for job creation, wood burners and reptilian wellbeing.

As I was sticking a stamp on a letter earlier I noticed the name of Franklin and a 350 yr celebration of electricity symbolized by a cloud of lightning flashes against a dark, night sky. It caused me to think how much has changed in our world since we have lit it up electrically. How much we fear the dark, how enslaved we are to pay the bills and wire our houses. Gone are the hearths and the soft candlelight.

※

This is the second of the most glorious days. Clear starry nights that have left a strong frost in their wake are surely melted by the strong sun bearing the promise of warmth

The Wave Hub Chronicles

and spring. Standing right on line between past and future, frost and fire balance in the eternal now. The stillness and lack of breeze gives access to the magic. Nothing is in the way of this most awesome beauty.

Walking around with the sun blinding the estuary to a mirror, dark blue shadows protect frost. The diggers who've been taking the sand to sell are now on the opposite bank of the estuary moving sand from a stockpiled heap over there.

We came upon the log on which the goose barnacle colony had lived, there were just a few remnants of shell. That's all that was left of the teeming writhing civilization.

A few yards away lies the dead seal corpse.

He is slowly becoming one with the sand, his flesh dissolving with gravity to return to ground. I stood by him with this very reassuring feeling washing over me. The seal slowly rots before my eyes under the bright sun's glare after enduring through the frost. Putrefaction appeared as a miracle, letting it all break down and rejoin the molecules of earth and sand that wait to reabsorb and envelope the carcass with total acceptance.

We have created so many scary images of half decaying skeletons and zombies to deliberately strike fear into our hearts and pump adrenaline through our veins, I wonder what purpose this really serves. Does anyone benefit by us fearing death so much?

This type of fear only serves to limit our experience of life.

I danced the Tai chi with my shadow sharply lit up on the sand, the sun directly at my back.

☼

Still the crisp bright weather holds as the seal melts into the beach, nasal cavities and spine appear as flesh leaves.

I have had news of a sudden death. I watched how this news affected me. I felt very far from home, adrift and unsettled, looking to the past and the familiarity of the already happened. Throw in a pebble and out it all ripples.

  There is very little progress over the back other than rearrangement of sand, mud and gravel for uncertain reasons. I am digging the potato patch just up by the brow of the bank so I can peep over and check what's going on at regular intervals. This is my net curtain point of contact now all is quiet on the beach.

☼

  How to describe the intense blues of the ocean's colours around the Cornish coast?

  We walked through the wreckage of decades of tin and copper mine workings where the streams ran green down to the pitted cliffs and white powder collected around the rocks.

  Submarine mining tunnels run one and a half miles out under the sea. They say that in a storm down there you can hear boulders rumbling away above your head, like rocky thunder.

  Back home the occasional sound of a jack hammer breaks the seaside noise of gulls and waves. Glancing over from the dune top they seem to be preparing the ground works. The very tall wave hub sign has been replaced by a gaudy company logo advertising the building contractors and moved 100 metres or so towards the compound where the substation will be built.

  Standing on top of the back garden bank in the still icy wind, I watch the exploratory intrusions into the earth by men and machines. Three high vis fluorescent men climb the bank.

  "Good Morning" I say, holding on to the washing line.

  "Is this the right way to the beach?" They ask.

  "Yes you may go through our garden," I reply making sure they realise the favour," where have you lot come from?"

  "Birmingham, where it was −9 this morning"

"And Dorset, but we don't want to trouble you…"

"It's no bother" I show them the short cut to the beach.

"What's your job then?" I ask nosily, looking for my reward.

"We're electrical engineers." They answer politely, then go on their way.

***

The bright spell of cold weather is persisting. How quickly we get used to something steady and ongoing. I'm no longer marveling at the icicles on the towans nor at the heat of the sun later in the morning making them drip. I've been going out with woolly hat, 3 thick layers, double socks and gloves for my early morning Tai chi session then on my return feeling madly overdressed as the warm sun broils me in my woollies. It's like living through a few seasons in a couple of short hours.

This morning I had a rare open door glimpse of very smooth flowing chi and felt infinitely flexible. Very often practice is simply that, practice: an endeavour, a discipline which offers a form through which this awareness may be experienced, attention permitting.

Occasionally precious jewels glint gloriously in the daily crown of the discipline.

I thought of Tesla and the electrical engineers and the power of the waves and me here going through my daily motions, plugging into the blue and the gold, the black rocks, the winds, sky and ocean. I bow to the four directions and feel an ache in my back as I stretch around my kidneys where resides my allotted portion of chi, bequeathed by my ancestors to live out my purpose in this vast continuum.

I had awoken with these simple thoughts. If we don't make anything, we limit our understanding of the creative process we are a part of and if we don't plant seeds to grow our food we don't really know what we are eating nor the magical reality that affords us nourishment. This ignorance is really a type of poverty and deprivation. It's no use to be watching too much telly in our real lives.

There are three men drilling with a big yellow drill into the earth. The drill bit is surrounded by a cage and powered by a generator. There's a hosepipe poked down the hole the drill is making. That leads to a square plastic water container sitting in a grill of metal. This arrangement reminds me of the dentists. They have that miniature vacuum cleaner to take away all the tooth debris. I watch the men about their earthen dentistry and have no idea what they are doing in the gummy mud.

Still this fabulous spell of weather continues crisp, cold and bright. Warm in the sun and cold in the wind.

Beyond the decomposing seal I find a freshly dead little porpoise, its eyes still intact although sausagey intestines were spilling out from a pecked hole onto sand peppered with the claws tracks of a hungry bird.

A cold wind sharpens up, bringing gathering clouds with it but not yet rain. The seasons are battling: life wants to thrust upwards but the fierce cold discourages tender shoots from emerging too far.

They have dug a big square hole, foundation depth and the jack hammers are banging away, the noise of them muted by the sand bank but still clearly audible. It all seems drab, an everyday sort of job, contemptibly familiar, with none of the challenge of collapsing dunes or devastating high tides. Perhaps the electrical part of it has

The Wave Hub Chronicles

its own sparky challenges but I won't get the chance to see any of that randomly happening as I pass by.

My enthusiasm for the project wanes as my craving for entertainment goes unsatisfied.

Someone has come and removed the seal from the beach. There were tyre tracks up to his body and then an absence, the corpse removed.

The little porpoise had been flapping in the waves, its body lulled by the motion, its motionless weight revived momentarily by the dance of the sea until left beached again, returned to stillness, silent on the sand, disturbed only by the pecking, cawing and hopping of the carrion ravens.

They are digging out the sand from over the other side of the estuary. This constant act of theft defies all reasonable propriety and casts an appalling spell of disempowerment over us as we watch the piles accumulating only to be driven off in trucks to who knows where for what purpose.

I am caught in an indecisive thrall, wanting to protect this precious, golden commodity which has taken its thousands of years to create and not being sure it is my place or business, this is surely the resigned position taken by so many of us after generations of learning the corruptions of ownership, the laws protecting the property of the law creators. What if our anarchist relations have a point when they say that property is theft? What if the indigenous people of so many lands are the truly wise and clever ones, knowing that ownership of the land, earth and waters is a stupendously idiotic and even dangerous notion?

How much back peddling are we now prepared to make as our greedy history proves itself unsustainable? We can't take it with us because it was never ours, so what exactly are these delusions we maintain and defend? The pressure is on us to realize a fundamental misconception: that it is us who belong to the world and not the other way around.

The Wave Hub Chronicles

The foundation hole is getting bigger as seen on this misty, moisty morning. The digger's got down to darker grit and stones. The two posts marking the ends of the pipes stand a few feet above and beyond the pit and opposite are another few posts marking something else I don't know about.

The piles of sand are growing down on the quay side as is the local indignation at the digger with its dumper truck sidekick fetching more and more as if stocks are running out. Its said the license from the Council to take the sand is due to run out any moment so the daylight robbery continues until the curtain drops on its legality.

※

Still heavy clouds hang low over St.Ives and Carbis Bay. The plants on the Towans breathe a refreshing sigh of relief following yesterdays rain. The beach is still and fresh. I spy a flash of high vis wear out of the corner of my eye as I move through my Tai chi sequence, drinking the damp air, feeling the love of the place and peaceful in my heart.

※

We are just past been the spring equinox and ever since the weather has been moist and significantly warmer. The Sea though still icy cold might soon be bearable enough for a swim in just skin and it's definitely time to start planting now all that frost is past.

The Wave Hub goings on are as dull as ditchwater. Watching the building site, they dig the foundations. There's no activity on the beach, no human contact, so no stories or chat.

The catalyst on the doorstep is the stockpiling of the sand from the estuary, which coincides directly with the shocking decrease on the beach as they dig dig dig away. The five year license comes up for renewal in May. The acceptable cover story is that they need the revenue from the sand to keep dredging the channel for the fishing boats. It just feels like the ongoing rape and pillage of the world and we're all so complacently plastic wrapper fed and centrally heated cocooned we fail to recognize the plunder or know how to act appropriately in response. I don't want to lie down in front of a digger driven by a man (probably) who's doing his best to provide for his family. I don't want the war, the fight, the polarized opinions. They just make everything seesaw, but how can I simply let this be? I know our mother Earth can shudder and we will fall, I'm horrified by our disrespect and this indignity.

This morning a swan was behaving very strangely. It sat on land and let the dogs approach with very little attempt to see them off save a sultry flap of majestic wing, as if it simply did not care. Its neck and head were discoloured, a mucky off white. I called the dogs away and thought about the swans being protected by the crown and the dogs being hung for treason, or would it be me?

The recent full moon has brought extremely high spring tides and wild blowing winds. Once again the beach has been ravaged by the wild sweep of water and wind, sand blasted and leveled, reformed and slapped about.

The loss of sand from this beach that I recently found out was called Avalon, continues to be shocking. A new swathe of rock has been revealed and the sharp, steep bank of sand has receded so far back that the path down from the holiday chalet area will soon meet with a massive impassable drop. The holiday makers have started to arrive as its Easter this coming weekend. This morning a couple of men with a plump and slightly arthritic dog were caught up on the bank with only steep access to the beach below. The dog slid down the sloping sand to play with some other dogs whilst the owners, apparently incapable of the descent were left stranded on the sandy stage calling for their pet to climb back up. She made a couple of half hearted attempts then gathered all her muster and valiantly struggled almost all the way to the top before sliding back down again.

I started to feel a strange sense of responsibility as if it were somehow my duty to help the overweight mutt be reunited with its owners but then saw that the younger of the two looked easily able to help with the predicament.

Maybe I feel a bit of hospitality is in order.

" Welcome to our beach, How can we help?"

Except of course, they didn't ask and who am I to assume any kind of propriety over their discovery of the natural mysteries of the environment. They won't want me impinging on their holiday experience. Nevertheless I couldn't shift a strange sense of the hostess thing as I went through my tai chi form.

A family of three, mother, father and daughter appeared from around the rocks as the tide receded and proceeded to exist within their beach holiday bubble, filling their yellow bucket with beach treasure just a little too close to me for it to be easy to ignore them. Spatial awareness being so relative and I having had the winter of extensive beach space we were presumably perceiving the distance between us differently. In town or city terms they were leaving me a wide girth as I tai chied practically on top of them in mine. I guess their filtering systems are switched up much higher perhaps due to the urban experience, but the little girl did kept turning her head to make contact until checked by her mum to look the other way.

No news to report from over the back, the ground works are going along as planned I suppose. Three separate holes are being attended to with poles, concrete and that rusty iron mesh that puts structure into slop. It's too early for me to describe what they're doing except as unintriguing. When the dog ban starts up and we're not allowed on the beach much we'll be walking over the back more often so I'll try and be more observant.

The Wave Hub Chronicles

April

Since the clocks have sprung forward the evenings are lighter later which gives more time for nosing about after working hours. The shift of daylight saving becomes increasingly disorientating as I get older. I find it now takes some time to acclimatize, jetlagged without the travel. Either I'm less flexible or more sensitive or perhaps a bit of both but I do wish they wouldn't mess about with the time. It's enough that clock time is an enforced concept as it is without then forcing these shifts. Best not to pay it too much heed, let the turning of the tides and the waxing and waning of the moon measure the moments. Appointments with clocktime can be just exactly that.

☼

It's Easter weekend and a bright and sparkly morning. Having rained all night the morning is rinsed and shines. It's the time of the cuttlefish. They are all washed up on the high tide line, all shapes, all sizes, busted up and in pristine condition, momentarily I wish I had a budgie.

Families are out on the beach to enjoy their holiday freedom. There's a man with a leaping yapping dog by the rocks and we chat above the canine noise.

"I'm interested in this rock," he says, "haven't been here for months but this is all new. I'm most interested in all the minerals and this is all fresh." He point to all the sparkling accumulation that is now exposed.

"What type of rock is it?" I wonder.

"Just common rock, shale, but it's the minerals..." And he gets out a little magnifying glass and peers through.

"I think I'll hack some off." He says as if to himself.

I head around the corner to my Tai Chi spot and while I do my form he chips away, the dogs bark and the water laps as the tide turns.

☼

The big waves are back, Surf's up. Great big fresh blue blades with curling white tops cut right up along the beach as the high tide rolls in. Up in the sky clouds scud along in time with the waves. Everything is moving together.

I walked fast the full three mile length of the beach, pushing my strides against the wind feeling the full motion of everything around me: spring pulsing forth into becoming, ocean curling huge, spray filling the air with damp moisture.

A grey mist sweeps in and soaks us all.

Holiday makers dribble off the beach into nearby parked cars, all efforts abandoned. Sticking close to each other, bands of family members walk together, roped with invisible genetic bonds, maintaining proximity in unknown surroundings. Even in, or perhaps because of the huge wide beach, the choice is to tag along behind each other in a line, following.

The dogs startled an Indian couple. They had just driven seven hours from Luton to come and stay in a seaside caravan. Her arthritis made it difficult to get down the sandy bank, he was so glad to be out by the sea, letting the grey, urban, premature ageing flow out into the salty air.

May

We have been away almost as far north on this land as you can go.

I feel discouraged on return to find the substation's development as boring as ever. Still in its foundation stages it's all concrete squares and regulation measurements. A road is also being built up to it from the gateway off the quay, a distance of 200m at most I would guess. Never mind that the quayside road itself is pitted and rutted to hell and back with bottomless potholes and complaints from all passing vehicles as they creep along to avoid mishap to their undercarriages. Some of us more reckless types with older cars and the drive them into the ground mentality try different approaches, such as speed, to fly over the crevices with alacrity. A smear of tarmac butter over the track's bread would smooth out the passage, but we'll see, I'm sure there are plans afoot.

I met Patrick on the beach later and he asked me had I seen the men in the dinghy with a computer, sailing back and forth across the estuary feeding data into cyber space?

I hadn't but had seen him chatting to a young couple in fluorescents with small backpacks and a single wheel, was it a unicycle? Did he know what they were doing?

He didn't, saying merely that they were lost.

The Wave Hub Chronicles

Just as I was leaving the beach the couple reappeared with their wheel, this time easily recognizable as a surveyor's measure.

"Are you part of the wave hub project?" I asked.

"Yes," they said, "we're down from Chester, subcontracted by the marine engineering firm to measure the beach for low tide depths. We thought we'd find a concrete plinth or something to show where we're supposed to measure from, we're not sure this is the right beach."

I showed them where the pipes were buried and enjoyed a brief moment of knowing more than they did until the reflection of local busybody came glancing off their expressions, so I left them pushing their wheel out towards the retreating tide ...

※

Watching the breeze blocks get stacked upon each other, grey cuboids becoming recognizable walls, I hear murmurs of snatched conversations between builders but little else.

I glimpsed the young beach surveyors all bundled against the cold north wind, walking around with their eyes seeing distance, depth and viability, freezing fingers jot down notes as the lifeguards' red flags flutter warnings that this is not a beach to swim from, the current is deceptive right next to the estuary.

The Cornish holiday season has begun. Can't walk here with dogs between these hours, can't swim here unless it's between these flags. And if you leave your car without a ticket (or sometimes even with one) , even on a piece of old rough ground that's knackered your car to get to they'll fine you or clamp your wheel until a local crook comes along to release you on payment of an exorbitant extortion of a ransom. It really is all about money and pirates.

The beach cleaners have been out, the stolen sand still stands in accusing piles across the river. The parking ticket machine has been emptied in readiness for all those fresh, grubby coins.

The Wave Hub Chronicles

Still the north wind brings a cutting cold that challenges the spring's green blooming.

Flowers ignore the chill and spot the green with stars of colour, shiny jewels in an emerald background.

Life pours forth to be met with a bright if chilly reception and the opportunities for gain from seaside attractions crank into action.

Postcards, ice cream, holiday chalets, cream teas, indoor attractions, outdoor activities, keep us busy as our people vote for indecision and a break from the old two way bickering, it's Election Time!

Doubt has crept in, surely an accurate reflection of our current concerns, well mine at least.

Who are we and what the hell are we actually doing?

We will all get felled at one time or another by the devastating effects of loss and death that poke and prod at the most ordinary or privileged of lives, irrespective of power, money, position, anonymity. Doesn't everyone have cause to question the apparent futility of their lives at some time or another?

So many of us function in clouds of confusion far away from that which makes us feel truly alive.

Are not thousands of people in our country prescribed anti depressants?

How many aisles of supermarket alcohol does it take to soften the edge?

It is rare to meet folk who are fully satisfied with their purpose.

I stood behind an Asian man at a supermarket check out. The white man working the till was a similar age. These two men's antipathy for each other was palpable.

The resentment at having to serve, the surly roll of notes unpeeled with disdain, placed on the counter rather than Into the offered hand, The responding change was slapped down with an audible mutter. These mutual gestures of dislike ricocheted between them despite their potentially similar experience of fundamental dissatisfaction.

I in turn received little other than the bare minimum attention at the till, but the little girl in the family behind me in the queue received a huge blast of loving affection, like a breathe held just too long and let out in a gasp.

※

I lost 10p in a bet that they would not yet be putting the roof rafters on the substation, now they've completed the walls.

Thinking of how long it has taken them to get this far I rashly risked the wager, believing it would be ages before they got on with the woodwork. But no, it seems spring has sprung and the urgent momentum for growth pushes the chippies on.

I won my money back later however when I bet another silver coin on the unlikliehood of they having put a waterproof membrane on the roof.

This yielded positive results and now they are concentrating on the place where the ducts reach into the compound. They have dug a sort of cellar and lined it with a greenish material.

I met and greeted one of the work force this morning as I walked with the dogs through the back stage scenery, now the beach is officially a holiday maker zone.

All my wave Hub info today has come from "The West Briton" newspaper in which there was an article and photo featuring the hub itself and a gargantuan coil of copper cable that is being manufactured in a Hartlepool factory, "Made for a job just like this".

Another article expressed concern about a new set of traffic lights that are likely to affect local businesses. The lights are intended to accommodate the extra traffic generated by the Wave Hub Renewables Park over a new bridge planned for North Quay. The Mayor is quoted expressing diplomatic concern and I happened to meet him a little later on outside the chalet, walking his fine dog.

He told me that the council has awarded £5 million to do up North Quay so that it can shift from the post Industrial potholed hinterland towards tarmac, employment and industrial units and perhaps even a waterside terrace.

A recently burnt out car would definitely be out of place in these upgraded conditions. As it was, it completed the warzone image perfectly as the north Wind blew sand and grit from the nearby aggregate yard into the faces of valiant passersby. So the big Dutch bank has now been given more public money to further their cause and help reshape the quayside infrastructure. How strange it is that this wealthy company should be thus supported from our pocket ...

It is hot.

Properly summer hot.

The beach is full of people sunning themselves, squealing in the icy, turquoise water, riding the waves, watching each others' bodies.

Over the back they have dug a great dark hole near to the green cellar and sunk in two wide bits of black plastic ducting that now stick up vertically, proud and erect.

They filled in around them and they appear to be providing ventilation for something.

The heat has turned everything into Holiday mood.

The switch has been flipped.

Bodies basting on the beach turn lobster sore red and the heat relaxes the world weary to recline on the sand.

Looking at the stripey windbreaks and detritus of drinks cans, plastic bottles, crisp and biscuit wrappers, it's hard to remember the winter's digging adventure on the beach.

Spirits are high as I swim around a newly formed high tide pool with my neighbour, enjoying encouraging wary

The Wave Hub Chronicles

newcomers into the freezing water just as we'd been encouraged in earlier by a woman brazenly claiming she'd been swimming for 40 minutes. We shivered and squealed making faces as the cold reached vital bits with its icy fingers.

Returning home I wanted to photograph the scene for a possible painting, the colours were so intense and vibrant in the sun's rays.

On the cliff top with my camera I scanned the scene.

A teenage girl shouted nastily up at me, accusing me of taking her photo like a perverted voyeur.

I marched down to her. She was the queen bee of the gang, assuredly posing in her circle. I didn't want to shame her but I felt furious at the accusation of sexual perversion and the infringement of my peaceful liberty that has us culturally looping the loop to avoid almost everything lest we are somehow liable for an injustice to someone : so I tried to explain. The combined shock of being directly addressed and the interest in painting of one of the lads in the group diffused the conflicting energies but I was left feeling sorely sad at the warped perspective so ridiculously over blown between adult and adolescent.

As I got back to the chalet a man and his wife were taking turns to pose beside our collection of sea buoys. I offered to take a photo of them both as if to exorcise the previous moment's ghosts. The man had both legs amputated above the knee and unusual mouth, nose and hands. After some talking it turned out that three years earlier he had been taken suddenly ill with an infection that had put him in a coma for two months and resulted in the loss of his legs and some internal organs.

He was so determined to live it was extraordinary. And so was his wife, extraordinary in her love that is.

※○※

The slates are on the substation roof, the diggers are working on the road and a couple of surveyors have been plotting the North quay characteristics and measurements. It all needs renovating, refurbishment and a general going over. We're in the last stages of decline, waiting to emerge from our chrysalis, butterflies in a new and polished era, former glory flying to the future.

※

It has been amazing growing weather, huge down pours followed by a riot of sunshine pulsing greenness forth, flowering into the world. I've been out picking elderflowers for champagne and cordial and honeysuckle for scent and beauty over the back behind the transformer. I hadn't started out that way as a digger and dumper truck were busy on my usual path but I figured they might be on a tea break as I returned and was right: the vehicles were parked blocking the path, abandoned for refreshment, fluorescent jackets strewn over the driver's seat.

The narrow path is being scraped wider, tearing brambles and evening primrose out from their sandy nooks. From the mid point on the path a plastic drainpipe pokes out containing wires, connecting something to something else.

For some weeks now a miniature stone circle with adjoining forest and quoit has sat by this path, created by a mystery artist. I kept meaning to take photos but hadn't and now one of the digger's claws lies casually amidst its ruins.

Don't delay with your intentions or they may never happen.

I noticed the digger had also driven right over the reptile fence.

Before it was sagging, now it was entirely crumpled and squashed to the ground, mocking its own inconsequence.

The slates are on the roof and beside it a Rachel Whiteread style concrete inversion sits where the pipes surface from under the dune. The roadway is becoming evermore established and the atmosphere of the place is beginning to change from wasteland to twenty first century smart. The eyes and motorcars of the tidy will no longer suffer the potholes, scrap and muck of outmoded fortunes, all will be transformed at the transformer with tarmac, shiny signs and neat buildings heralding the way forward to sustainable electricity sources, via the grid, of course.

Out walking the Tinners' way on Penwith moors recently, we came past the most beautiful wind turbine work of art in a garden by a little cottage. Designed by a woman, built by a man it sits atop a small round building covered with slate. A vertically tapered wheel turns slowly with the speed and grace of an old waterwheel.

Because it isn't certified by experts or made out of regulation materials it cannot be connected up to the grid, still it powers the needs of the house and the efficiency and elegance of the design charges the environment with its magnificence.

Save Our Sand.

A local pressure group to do just that had a meeting the other night. I heard from a neighbour that a new harbour master has been appointed, who not only seems to be properly qualified but as part of the review of the job in hand, has suspended all dredging and excavating of sand from the estuary and thereabouts, pending evidence that it is indeed necessary for the running of the harbour and not just for making a quick buck.

Signs prohibiting kite surfing on the beach have also gone, thereby allowing kite surfers to pursue this leisure activity legally rather than make exhibitionist criminals of themselves every time the wind and waves combine preferably. This harbour master is also a marine lawyer so it may be that he knows something about something, but whatever that might be at least the sand is lying safely for the next little bit of time. The bottom line on this issue lies with the Council who grant the authority to dredge and sell.

The Wave Hub Chronicles

June

The summer's heat is steadily building day upon blue day increasing in light and warmth. It feels like years since we've had such a lovely spell and it is delightful: Clear, breezy, warm, turquoise and gold, the days culminate in the massive red ball gently falling into the sea behind the island at St. Ives.

At this time of year it sets almost directly opposite the chalet. Glorious colours invade the skies tingeing everything pink, peachy golden and sublime, the water becoming liquid silver, bronze and molten gold, the turquoise shifting into black and back, as light shimmers indescribably on its surface reflecting upwards following the undulating energy, then finally leaving pink puddles in the velvet blue twilight.

Temperatures in the water have varied wildly. For a few days it seemed to be warming up, then suddenly the sheer iciness would have me squealing. I'm convinced there's something or one pouring ice cubes into the ocean that are ending up dissolved but not forgotten on our beach. Perhaps it is the very same polar ice cap melt water lapping on our shores, for where does water stop and start? Is the Gulf Stream getting pushed as I write, as you read, right now not in some mythical future? It certainly feels like it.

The Wave Hub Chronicles

Of course the river water from the estuary chills it down some.

Miles up the coast the sea water on the Hartland peninsula in North Devon is definitely warmer than down here. The subtropical flora of Cornwall has depended on the warmth of the gulf stream's pathway to survive and thrive. The magnificent palms, which have been in phenomenal flowering mode this season are a prime example. These exotic beings stand so tall and straight up all around West Penwith and are blooming madly, their flowers an explosive bauble, a static firework wafting a rich sweet aroma all around themselves.

We have now just passed the longest day and the blessing of the sunshine seems all the more gracious in its largesse.

Basking lazily in the Mediterranean heat the world seems eternally benign and easy going, the winter's icy winds an imponderable memory.

This morning the Substation development was awash with activity, a hive of mechanical buzzing: state of the art diggers and cranes with hydroscopic arms and insect like attachments for executing any challenge were swarming around shoveling gravel onto a meshy scrim laid around the station building which now has been fitted a door. They dig trenches then drive up and down the new bit of road way.

We passed by on our way to visit the new baby buzzard chick in its nest just a couple of hundred metres away beyond the (thankfully) restored reptile fence.

No sighting today, but the nest itself looks magnificent perched in its eerie of Monterey pine branches, a beautiful construction which looks through the binoculars to have something white and fluffy in it like a chick down duvet. All eight sea gull chicks are present and correct, in residence on the old power station roof with a bird's eye view of the proceedings.

The Wave Hub Chronicles

Their strange spotty heads are difficult to imagine fully grown into the sharp beady eyed adult birds that stalk unwary tourists and their unguarded pasties and ice creams.

Returning back via the track to the car park and the former disabled parking lot that's now the works access, big machinery trundled by on caterpillar tracks heading for the estuary and along onto the beach proper.

At high tide the early afternoon's heat called for a refreshing dip. The beach was transformed. An area was cordoned off with holey orange plastic, staked out like a useless windbreak down two parallel lines dividing an enormous crane of a digger, an automated monster machine, a few guys in hard hats and a load of long ridged metal girders from the vulnerable limbs and torsos of the holiday makers on the beach with their stripey primary coloured trunks and bikinis.

What a sight as the machine maneouvered one of the massive girders into place with its hi-tech multiple gripping options to pile drive it in the sand alongside a few already placed at the high tide line. Presumably this is in preparation for the cable coming in from sea to shore. I wonder innocently whether they have taken into account the variable height of the high tide.

November's high tide seas made so much mincemeat of long days careful preparation...

All that aside, this was really very entertaining. Large scale beach works right along with children with buckets and spades, women with suntan lotion and radios, men with newspapers, couples peeking out from beach shelters to check on their kids digging in the sand, until all of a sudden an ear splitting noise of grinding machinery, metal grazing on rock and drills screeching blasted the beach in such a way as to make it absolutely impossible to relax. I felt for the people whose hard earned holiday was being penetrated by the audible violence that ricocheted cove to dune to ear as I swam about in the blissfully surreal and sometimes icy waters.

The Wave Hub Chronicles

As I took the dogs round by the estuary early this morning I was forced to grab the one that chases cars by the collar, as two large moving bits of machinery trundled up from the beach via the works access just ahead of us. There was one with a big telescopic arm and the other a tractor and huge flatbed trailer. They'd been delivering something down on the beach.

After Tai chi for me and rabbit hunting for the dogs we came back via the transformer and blooming sweet honeysuckle to see another hive of industry occurring in the compound there. A giant crane was lifting a heavy metal piece of jigsaw and fixing it in place in some kind of connect up that I couldn't decipher. In the background behind the hard hatted men stood two huge wagons with gert reels on the back of them.

Nine million quids worth of copper perchance?

The criminal part of my mind imagined sneaking around after dark and hauling off a few skeins from the massive circumference, nipping up to the scrap yard and getting a good price to see us through some summer fun. Then I remembered that I was a well brought up decent sort of a person not overly given to crime. Even so there is something about raw materials that is so properly valuable, muck and brass valuable like a wallet stuffed with notes, that it brings out the greed in me.

I rushed back to get a camera to snap some shots and I felt sure as I took the photos that I was being assessed by the workforce as a possible threat or thief. I'm sure they'll put security on watch over those gigantic reels of metal.

Later on one of the workers was passing the bottom of the garden as we sat after lunch sunning our lazy bones,

"You can take a short cut through here" I called out to him, ingratiating myself,

"How's it going down there?"

He was only too pleased to chat and grateful for the short cut.

They were putting in the girders so they could dig behind to create a junction box where the big cable from the sea would split into two and travel separately through the dunes to the substation. The junction box is to be set in resin and the metal girders removed so everything will eventually be hidden under the sand. The sea going cable is being transported by boat all the way round the coast from Hartlepool south down the East side and then along, sometime in July.

I'd love to go aboard.

Today at my high tide swim the digger was sitting on top of a pile of sand shifting long girders from horizontal to vertical then burying them into the sand and rock. Not very fast work as each bit takes time. The guys working looked happy to be in such a location for such a job: they are simply a construction company involved in a variety of building projects so this one is pleasantly unique. The clement weather and seaside atmosphere makes the work look playful and intriguing but there was hardly anyone out watching as England were playing Slovenia in the world cup and the rest of the beach was deserted except for a couple of other bathers bobbing in the chilly but charming brine.

※

News from over the back.

The fluffy white stuff in the buzzards' nest is in fact a large chick. It's so fluffy and huge it takes up the whole space in the nest, perched there among the pines. What we thought might have been a snaffled duvet, now clearly moving, is a big baby.

The new transformer is in place upon the concrete plinth, the jigsaw puzzle completed upon the inverted house. It sits, brand spanking new above the holes for the cables, ready and waiting for them to come poking through and get wired up like a plug.

Meanwhile back at the beach there's another opportunity to wonder whether the contractors have a tide timetable with variable water heights as they seem to be getting repeatedly caught out by the tide rising higher than yesterday as if they might not be aware that it changes.

Today the great big crane with the caterpillar tracks and grippy arms was sitting on top of a mound of sand as the waves washed over the waiting metal lengths (playground slide in shape) and undermined the mound's edges until it looked in danger of toppling over. The driver quickly drove it down and round out of the waves' way, the narrow strip of beach narrowing in retreat from the water. Within the waves the water was lively, full and buoyant, the recently abundant jellyfish were less plentiful than in these previous days when the water was so full of the bluey purplish ones that whilst swimming you had to pick your way past them and their nettle like sting.

There have recently been reports of the Lion's mane jelly fish in the Lands' End area which has a dangerously paralyzing sting. These here are relatively harmless as I reassured some holiday makers, although I find it difficult myself, not to be put off by their strange and gelatinous being.

It is clear now that they are constructing another coffer on the beach, uncannily resembling the other one, almost identical except in a very slightly adjusted position.

※

It's all going on.

Summer in full flight: Long days of intense heat, dry and scorching. A heady build up of temperature tension, the air is thick with sticky challenge and we begin to anticipate a storm. Gusts of wind flap about distracting the intensity, then quietly in the short night the rain pitters on the

The Wave Hub Chronicles

grateful ground just enough to promise a quenching. A day or so later it rains, purely and simply wet.

Rain, filling the water butts, dousing the plants, washing and rinsing the air, fluid and inimitably refreshing, something to be grateful for.

All the while the baby buzzard grows in its nest. Standing up now to pull apart baby bunny flesh, its legs look powerful and predatory even in their fluffy knickers. Mother and Father Buzzard fly to safe distance at any approach to the nest in an effort to distract. They sit on trees, flagpoles, telegraph posts watching over their baby, their equivalent distance from living room to nursery for their flight is smooth and fast except perhaps when mobbed by gulls and crows.

The gull babies are on the roof of the power station and becoming sleek as they prepare to fly someday soon. They are losing their gawky chick look and gaining the fierce beadiness of their parents who swoop and caw as the dogs run by obsessed with chasing bunnies.

The gull parents aren't taking any chances or maybe they just like having a go, venting a bit of gull aggression. What if their thieving pasty and ice cream snatching habits has caused a dietary imbalance and all those processed foods create addictive and peculiar behaviour? After all it has happened with us humans.

Everywhere there are flowers, beautifully scented flowers. Honeysuckle and privet abound in the hedges along with vetches, umbilliferae and lots I don't know the names for.

The privet scent jogs a primary school memory evoking random snippets of memorable imagery: a single storey red brick building, an advertising billboard with nicotine stained fingers and a nail brush, a caution that you couldn't scrub the inside of your lungs, double decker red buses seen stopping over the top of the hedge before disappearing under the flyover to their next stop, I'm amazed at the dexterity of mind that can travel back through decades at the smell of a bloom.

The Wave Hub Chronicles

Flowers.

Making space for each other as they succeed one another, some intent on strangulation: convolvulus trumpets its competitive beauty, others mature with their pips and fruit just starting to show, pregnant bulges underneath spent petals.

Now past the solstice, summer's longest day no longer offers future promise, it's all being delivered.

On the beach our metal coffer is complete and in place, a defense against the wet waves, a dry place to make electrical connections.

The engineer in charge is a friendly man. I chat to him as I emerge from a high tide swim. He is a Cornishman who knows the sea, a diver and boatman. I tell him about some of the apparent ignorance of the plans, he laughs and tells me stories of other jobs where they wouldn't be told and had to find out the hard way, the intransigence of projects and people used to making things happen because they say so.

Every sort of big machine is on the beach to facilitate digging, pumping out and drying. Nightly about two feet of water accumulates unavoidably inside the coffer as anyone who has ever dug in sand will understand and short of digging deeper and slapping loads of concrete down, this is how it will carry on until the connection is made, the resin is set and the whole thing buried.

He is dubious that the cable will remain buried given the ever shifting sands and I agree. We are after all dealing with a living world, on the one hand we want to take power from it on the other we still insist that it keep still whilst we faff about thinking this should work. The digger, cherry picker, generator, pumps and various fixtures all lie about on the churned up sand as we sit on rusty metal girders, he drinking tea, I in my toweling post dip robe.

Later that evening at the second high tide I meet the young man who has been wild camping in the dunes, he is a bit pissed and comes in for a dip, lurching about as he

The Wave Hub Chronicles

tells me he is celebrating his 25th birthday. He's escaped from London and wants to work and settle in Cornwall but can't find a job. I wonder what he's left behind and resist finding out. He seems perfectly capable, doesn't need the interference.

The engineer tells me that the ship to shore work is probably going on in August now. I hope I am here for it.

We laughed. No other project would be allowed to go on in high summer tourist season but because this is Hayle, this doesn't seem to be a consideration. The poor relation, the disrespected rubbish dump, how does it happen that certain places resonate with unworthiness?

Relatives who are coming to stay in a nearby chalet are concerned, Their Landlady has warned them that the beach has become an industrial zone. I suppose she may be fearful for her financial interests. I reassure them that if anything the works are enhancing a memory of the magnificent industrial glory of former times but otherwise relatively unobtrusive.

The beautiful evening primrose is the queen of the towans. Everywhere you look she is flowering with her lemon yellow bell. Even inside the metal fence of the transformer, where gravel ground restricts almost all other growth, she blossoms. My friend suggests she is transforming negative energy. Whatever her remit there is power and strength in this plant so prolific and extravagant in our neighbourhood.

They have unwound the cable off the big reels and poked it through the black pipes to await connection on the beach. It was exciting in an unexciting way to see the huge crane dandling the massive reel in midair so the cable could be fed through the eye of the needle using Stone Age planking technology to assist its passage.

To be precise there are two different cables: one black, the other black with an adder like yellow stripe running its length. They are now busy burying them again.

The developments at this stage are so small and incremental it would be easy enough not to notice until later, at some indistinct point in the future something new appears to suddenly exist.

The Wave Hub Chronicles

July

The rusting iron coffer sits in the sea at high tide. The engineers have attached some meccano like shelves to either long side and winched up generators to sit out of water's way. They will power the pumps. Hoses are slung bedraggled over the high sides waiting to belch out unwanted wetness. The shelved generators look perilously close to getting a proper soaking should a wind get up behind a high tide. So far they have remained dry enough to function. The whole arrangement sits now waiting for the boat to arrive with its cargo of ten nautical miles of copper.

※

In true small town style we heard the latest wave hub news from another neighbour. He lives in a state of the art, newly built chalet which the harbour company were obliged to provide for him.

Prior to this he had been living in a block of disused and abandoned public toilets on South Quay for about 14 years when they came to evict him. Situated as he was, plum smack in the middle of their redevelopment plans, the harbour owners wanted him out of the way. Local friends championed his cause and discovered that he had squatter's rights and today he is what you might call a local celebrity.

He still spends his days out in all weathers sitting long hours at the town's boundary stream, relating with the flow.

We haven't missed the grand arrival of the cable, apparently , he tells me, its coming via the north of Scotland and down the west coast.

He used to deal a bit in scrap metal so we laughed about the value of this haul.

Perhaps the most precious bit of all was the fact that this neighbour was even talking with me: often I don't get as much as a bye nor leave and sometimes its wisest to get right out of the way to avoid an earful of profanities.

※

There was however, absolutely no sign of a boat as I went down for my high tide swim. It was damp and misty, it seems ages since our sunny summer spell. This damp drizzle feels well and truly set in, the sun a distant memory and that promise of a long, hot summer just another reason to grunt and sneer in true British derisory despair.

Upcountry we witnessed some beautiful straight down rain in lush greenery. There is a satisfaction with that colour green, it settles your heart to gaze upon it, knowing that all is well with that greenness, as it should be.

We walked among valleys, hills and woods soaking with falling down wetness, moisture clinging in swathes to branches and leaves, the air warm and humid, our very own rain forest.

Now, however, back by the coast, the part of me that's interested in self gratification is wanting sunshine and dry and the boat to come in before I'm due off on my travels again.

※

The Wave Hub Chronicles

Along the beach behind the coffer the path down from the holiday chalets has been shut. The erosion of the dunes has become so significant that the path now ends in a sharp and dangerous drop with the added danger of ancient rusty metal dumped decades ago reemerging through the sparsity of sand to poke its reminder, taunting tourism and holiday crowds with the shame of trashy history repeating itself.

If we're capable of joining the dots sufficiently it might be worth noticing that natural places need respecting and that digging tons of sand out from a place may leave a whole lot missing.

Whatever, Nature once again is showing herself infinitely powerful in the face of our little human ignorance.

It was some days later as we were walking up the coast path North to Hartland Point, when passing through a place with a mobile phone signal, as distinct from most of the wild heather and gorse strewn cliffs and coves, I received this text message from a friend who lives up the other end of the beach:

"They are unloading the Wave Hub right outside our door, well u know a bit out to sea."

I shivered with excitement and was glad to know we were due home later that day.

The Wave Hub Chronicles

Rusty Bedsprings by I P Knightly

August

When we did arrive home tired, smelly and ready to shelter from the penetrating mist, even a hot bath couldn't dissuade me from a quick peek over the edge of the dunes. Pushing through the soaking marram grass thinking of hot water I was miraculously restored and invigorated by the sight of a great big ship just a few hundred yards out to sea from the low tide mark. There were loads of hard hatted figures moving about on the beach with lots of different bits of machinery. A string of pink buoys bobbed about in a line out to sea, the sand was all churned up by the milling about.

Having soaked in the scene I returned to soak in the warmth of the bath then was out again armed with a camera and a brazen enthusiasm to find out exactly what was going on.

The first two fluorescent young men I met were climbing the dunes smoking fags and looking glum. They started to offload their disappointment before checking themselves and moving quickly away. The next few I approached simply said they weren't allowed to speak about it.

Undaunted I marched on snapping glimpses of the ship, the buoys, the coffer, machinery and the frankly damp squib of a scene, trying to understand the mood and possible causes.

I found two local volunteers who were only too willing to say how the marine engineers had wasted time (who along with tide wait for no man), cocked it up, not followed the tide and generally weren't particularly up to the mark. We had a very satisfying whinging agreement that some people couldn't relate very well with the movement of the sea when trying to get a job done.

These volunteers were supposed to be keeping beach users safe as the procedures happened but since the weather was so uninviting for holiday beach visitors and the procedure wasn't happening, it was safe to spend this time complaining. Thanking them for the information I walked on out toward the ship, still in best holiday mode and approached a guy in a hard hat sitting on a dumper truck parked by the end of a cable to which the line of pink buoys were attached. He too seemed only too glad of the diversion as his feet were soaking wet and had been since the morning when he'd had to stand in a deep puddle to cut some metal. He wanted to finish and take his boots off.

A fiasco, I think might have been his words but suffice to say the winch had broken, the cable had kinked, the tide was too low so they'd have to start work again at 3am.

The whole project was thoroughly expensive and would take years to pay off, same as the wind turbines which he also knew about and how they needed diesel engines to get them going, can you believe that and it's still years before the grid starts earning money for itself.

All this wet booted hearsay had me quietly wondering how the grid itself would profit and how effective dampness is at rising up the body to drown the spirit in negative conclusions, when he informed me that at high tide tomorrow morning they would be floating the cable in from the big ship.

This big ship named the Nordica, works eight months of the year laying pipes and cables in sea trenches and the other four months up in Norway and other icy places as an icebreaker. I looked over at the ship and fancied I could see a cold and icy aura surrounding the vessel in the mist.

"How many people are on there do you think?" I asked

"Ooh, loads" said the man on the truck, his long hair dangling into his baccy tin held in cold gloved hands at the end of bare tattooed arms, "They've got the divers on there all going in to guide the cable and dig the trench, they don't get paid much those commercial divers. On the oil rigs they can get 5 or 6 hundred a day but they'll be lucky to get that here in a month."

"Never!" I said, appalled at their impoverishment for seemingly dangerous work,

"Oh no, they're just diving around in the water, like going swimming" He said with an authority I somehow doubted.

"So what exactly are you doing here?" I asked

He was guarding the end of the cable for or from innocent members of the public passing by.

"Health and Safety gone beserk, you can't do anything in the building industry without health and safety."

"It's barmy," I chipped in wholeheartedly, "What about the reptile fence?"

He looked at me blankly and I watched him decide not to hear.

"How deep will the cable go?" I asked to salvage the conversation.

"2 metres under the sand to about so many kilometers out to sea and then along the sea bed."

"Oh." I replied then I started to tell him about the level of the beach dropping so dramatically.

Suddenly, something fell down in the vehicle's cab from the steering wheel and another guy came over on a quad bike, so I said goodbye and walked off in the direction of the sea and the boat.

On my way I met a man from Liverpool who has been holidaying here with his family for over 20 years.

The Wave Hub Chronicles

The Wave Hub Chronicles

"Have you any information?" He asked."I don't know anything about this." I told him everything I know including all the dumper truck man had just said about there being hordes of people here tomorrow to watch this historic occasion because it had been on Spotlight Southwest TV every night this week and the energy minister himself had stood on the beach last Wednesday giving an interview.

I told him I was following progress intently and was interested in our human need for sustainable energy.

I could feel my ego at that slightly anxious edge, wanting to have something to say that would be listened to and not wanting to think people knew it all because they'd watched the blinking telly with the politicians claiming the spotlight.

I heard myself sounding like the dumper truck man and understood the desperate thirst for empowerment that beleaguers us, my mind running through all of this in the time it took to tell the man from Liverpool,

"I don't have a telly"

He in turn told me that the 360 miles it took him to arrive here would get him to Fort William in the other direction, now here was a person I could reasonably converse with.

We talked with a bit of British pride about the technology involved in such an innovative global project, and how we've not got too much to be proud of otherwise.

"Oh no," he said, "especially not with what BP has done over in the Gulf of Mexico there."

We both made a face of mutual commiseration to be connected by nationality to that worst polluting offender of a petroleum company. But we still drive our cars...

Spent with chat we bade farewell and I took the last of my batteries' worth of photos of a child jumping in the waves with the Nordica in the background, mindful that it's not done to photograph kids without consent. But this one was a tiny silhouette with no detail in a large misty background, his parents standing by watching, so I figured no one here would mind or even care less what I think about the children and their children and their children's children in these uncertain times, as long as I'm not a member of a paedophile ring.

On my way back shorewards the dumper truck man had hitched up the cable to his tow hook and was dragging it back in towards the beach so he could get those dratted boots off.

The lethal curve of moving whipping cable went unprotected by all the pissed off workers and random passers by with the supreme irony that it was now extremely dangerous and could have sliced off a limb at speed.

Only one engineer was hopping about with an eye out, not perhaps so much for safety but because the kinking of the cable would have financial and functional implications.

The mood was low although the weather had lifted slightly.

The baby buzzard has flown its nest.

Up bright and early with the blue sky and the sun. I went to see if anything's happening and it's safe to say that nothing regarding the cable seems to have changed since yesterday despite them working on their mobile phones until midnight and starting again at four to catch the tide.

I keep meeting people with cameras walking away from the scene with that resigned look that is so uniquely British, it says I knew it wouldn't come to anything much in the end but I couldn't help hoping.

Three management types were stood above the beach in good humour, saying there was still something wrong with the pulling gear but that big ship was surely costing them the earth sat out there to sea, about 1.8 km to be precise.

One of the dogs had run onto the beach to play catch with a family sat on the shore. She is not allowed on the beach at this time but there's no way she could ignore this chance of family fun. As I was retrieving her they asked if I knew what time it would all be happening as they were waiting for the big moment,

"Better than those wind turbines, I wouldn't want to look at them."

I told them I thought it would more likely be tomorrow as the winching gear was still out of action.

Whether they believed me or not, I left them sat there waiting for the show as I dragged my dog away by the collar.

With some visiting friends we go out later in the rain over to where the action might be seen. There's still nothing doing.

Everyone is out asking what's going on, clutching cameras.

The TV vans are parked in the car park.

Just by the vans, overlooking the bay with the Nordica in the background, one of the bosses' publicists is being interviewed. Standing in earshot I hear

"1.8 km ... 2km deep and the Nordica ... winching cables ... historic moment."

He then gets asked to repeat all that in a slightly different way by the interviewer whose sense of self importance seems to glare out brilliantly from him as he valiantly deals with the incompetence of his subjects.

A woman from town starts to chat with me and we start to cackle. They ask us to keep our noise down so we talk loudly as we walk away, instantly transformed into naughty school girls.

We greet my unpredictable neighbour who doesn't think much of any of it.

I offer myself to be interviewed from a local resident's perspective, but the shiny interviewer refuses.

The Wave Hub Chronicles

A group of engineers walk past the chalet as I'm going in.

"Hello Gentlemen." I say and they seem to be pleased to be so addressed.

One of them seems a little bemused by my interest in the project and suggests that I've been at a bottle of gin. When I declare that I'm not a drinker he suggests it was a reefer. When I tell him that I don't do drugs he wonders what it is that is fuelling me. I suggest it might be the sea and not so dissimilar to what they're trying to capture from the wave hub. His expression confirms my lunacy in his eyes although his colleagues seem a bit embarrassed by his affrontery and inform me that 2 pm is the deadline for what I'm not sure and that they'll be drawing the cable in at this night times high tide.

Later on the Nordica is still waiting patiently as the pink buoys bob in alignment and the many project managers talk openly of a few more hitches, public criticism and the need for patience as they try these things out for the first time.

They tell me of their experiences on other jobs: arctic expeditions, gas and oil explorations as we while away the waiting friendly and companionable, various crowds dispersed over the sands.

A group of young people on holiday from Up North ask what's going on. I tell them the story so far as I know it and then a project manager who I've never seen before steps in to give his full and authorized version, so I leave him to it and carry on about the beach watching and listening.

Now it's tomorrow high tide and the Nordica needs to come in towards the shore a bit. They've laid a large concrete slab at the low tide point in line with the ship.

The Wave Hub Chronicles

The Nordica can hold its position without anchor and will keep within a 50 metre corridor for the whole 25km cable length out to where the wave hub is situated. (Or will be by the weekend when it's scheduled to be dropped off the boat.)

It was raining then clearing, then drizzling and finally at sunset high clouds dominated the sky.

High tide will be around 11am tomorrow morning, for floating the cable in and making the connection at the concrete slab.

※

Nothing seems to be happening this next morning on the beach but I meet the wave hub's manager as I'm putting out the rubbish and we chat about the stuff.

The TV vans have now been parked up in the car park so long its like they've moved in. I clocked the TV man noticing my conversation with the boss and felt a smug satisfaction. Maybe now he would understand I wasn't just a nobody fit only for turning on the telly to listen brain empty to the important information he alone knows how to transmit. What is this strange antipathy I feel?

Later the ships' lights burn like birthday cake candles as she waits another night out in the bay. Our visitors wonder why they don't just use a barge to bring in the cable.

※

Days pass and still no sign of anything progressing. The initial fanfare of anticipation has lost its trumpet, there's only so much delay that can hold our attention before we lose interest.

※

It's a sunny day and I meet the new security guard and a winch man on the beach. The guard is glad because he's got a job until at least Saturday's high tide. Standing all those hours on the beach is meaning he's sleeping well at night. The winch man definitely wants the work as he's off on holiday to the south of France.

I've been away for 16 days and return to find the cable still adrift and unlaid at sea.

I'm glad I was not here to witness the spirits dashed daily on the rocks of disappointment, I can imagine the stagnating wait as hope arose each dawn only to set unfulfilled each dusk.

But I'm happy that I'll get to see the cable floating in triumphantly as I'm certain it will sooner or later.

The workers on the beach and around are exuding a sad deflation, a lack of enthusiasm, a defensiveness.

I can't help but be aware of the singular lack of wisdom regarding our relationships with the world around us. I have seen no one pay their respects to the Sea and yet they are wanting full cooperation and having trouble getting it.

Walking on the sand this morning as a digger rumbled and screeched and a dumper truck scraped and ferried piles of disturbed sand from one place to another I felt cold, numb and saddened.

I heard from a friend who had been dining with the skipper of one of the support boats that they had simply failed to float the cable with the floatation aids they had prepared, the cable had been too heavy all 1400 tons of it and the copper cable had twisted and bucked and kinked as it rolled off the Nordica and despite some desperate efforts to strap more buoyancy devices to it (blow up lilos, inflatable killer sharks and dolphins grabbed from unsuspecting holiday makers) they failed to hold up the weight of the copper which bucked, hooped and sank.

So it had been back to the drawing board whilst the Nordica remained languishing in the bay at thousands of pounds per moment.

The Wave Hub Chronicles

The Wave Hub Chronicles

This morning I walked out along the beach feeling slow and displaced after a long night's dreaming of the life forces of the Earth. I didn't feel like chatting to the guys working on the beach. Instead I watched them. I quietly stood and watched them as they faffed about ostensibly leveling the beach, cordoning off an area where there was a pool of water with metal stakes and ticker tape and there they were again looking for all the world like a bunch of boys playing about with a back drop of tourists making merry in the first clear sunshine after days of murky sea mist and rain.

A young engineer came over to warn me that the sand might be slushy and prone to sinking.

I said thank you and that I knew the beach well enough to negotiate danger, then I asked him how it was going.

He was immediately defensive, obviously having had to explain the lack of job completion too many times already and started to blame the very specific requirements and conditions that hadn't as yet materialized to manifest success, no one really understood this and the media were piling on the pressure.

I assured him I was simply interested and tried to ask about the new floatation devices but it was clear I wouldn't get anywhere informative with this young man, pleasant though he was, for he was well schooled in the art of self preservation that we would probably all recognize, when attempting to defend the indefensible.

What he couldn't say outright was that Nature was getting the better of them, that it felt like a battle, that they were squirming under the gaze of the TV camera's microscope.

Perhaps I should have been more upfront about what I feel about deeply honouring the Sea herself.

Am I in fact really frightened of naming such a force?

Or speechless in front of such mystery?

The Wave Hub Chronicles

Or apparently stupid for not having clever words of explanation?

I walked up to the water's edge where the holiday makers were jumping in the waves, giggling with delight and screaming with pleasure and saw how deeply we crave the Ocean's blessing: a week's holiday at vast expense, a drop in the ocean of a working year.

Tired wage earners stand calf deep in the water looking grey and empty, hoping for the miraculous cleansing of the Sea's waters and air.

And they are given.

But we as a culture have no way of showing our gratitude, giving thanks, making offerings for being here and receiving this miracle. We act oblivious, with the notion that it is simply a resource that we can use up in the service of our little lives.

Returning beachward I saw the young man, smiled and said that I'd managed to safely negotiate the beach.

I felt sorry as we walked in our different directions as I had sounded smug and the distance between us had increased.

Another group of workmen were gathered by the cordoned off pool. Approaching cautiously, I asked them how it was going. I could smell their reluctance to talk, but this time I managed to allay the fear that I was critical of them, instead telling them what I knew of the beach and how it has been changing and what has been done to it. They relaxed a little and said they were planning to get the cable in during the next 24hrs, probably in the early morning.

It's forecast very heavy rain tonight but I didn't mention that as I guess they would already know. I wished them luck.

I want to make an offering to the Sea on behalf of this project because I feel that at least its trying to get something right in relation to the earth or I hope so anyway, we simply have too little understanding about relating appropriately to these incredible elemental forces.

This is a massive tragedy for us, the source of so much heartbreak. And we don't even really know that we don't know. We are so far away from letting ourselves appreciate the devastating isolation surrounding our lives. Small wonder I suppose, given the religious wreckage strewn through our recent human history.

So I made some offerings to the ocean as I swam at high tide and then the rains came heavily falling in torrential sheets that demanded a stopping and listening to.

This morning early I went out into the sunshiney bluster. The cable has successfully reached the shore lagged by continuous red inflatable cushions, industrial strength, strapped to every available bit of space with an occasional buoy gaffered on for good measure.

The cable lay innocent on the watery sand and I felt a surge of congratulations well up and out of my mouth to the assembled men working.

In response they warned me not to walk on the possibly sinking sands.

It seemed a little sad that they couldn't receive enthusiastic praise either because I'm not anyone who is officially anyone to give them it or that feeling the albeit momentary pleasure of job satisfaction is so rare when you are busy watching your back and failing at your peril.

This latest batch of engineers were all new on the project; brought in where others had indeed failed by underestimating the weight and scale of the cable.

75kg per metre for the conducting of as many as 33,000 kw of live electricity plus the carriage of two channels of delicate fibre optics for relaying all sorts of information from the sea bedridden Wave Hub to head office nerve centre made a heavier and more unwieldy challenge than they'd anticipated.

The fibre optic stuff is much less robust than the copper and would be seriously problematic if it ruptured anywhere along its 25km length.

The previous team that hadn't quite managed the job and have presumably in my mind been demoted or redeployed as yard sweepers once their knuckles recover from the rapping and the hot coal burns have cooled.

After all, the TV cameras had been steadfastedly fixed on the contractor's logo as the firm didn't deliver, day after day.

Laughing stock might have been the words but you can imagine the vicious carnage and stunted career prospects strewn all over the boardroom floor as project managers got their marching orders. It might explain why their colleagues didn't feel much joy in accomplishing this particular challenge because there for the grace of God go all of us. There is no real pleasure to be had from someone else's misfortune unless you're a proper sod.

They had seemed like pleasant enough people. At least they were friendly. This lot aren't at all approachable. They inherited the tension of the task and are oblivious to all else.

This created a kind of sulky feeling in me, which in turn gave rise to a "fuck you then" as I strolled in the open blasts of wind lit by the first really bright light for a few days.

Later, all such resentment, a curious memory, I bobbed up and down in the wavy water, recharging in the ebb and flow of the swirling mass.

※

The rain came down in sheets again today, steady and monotonous, intent on the laundering of all things grubby

The Wave Hub Chronicles

and soaking through all claims of waterproofability. We're so lucky with our skins, what a mess we'd be without them.

Totally drenched through I took a calculated gamble on no dog warden braving the monsoon, to stroll recklessly around the edge of the beach in the tipping wetness, picking up a couple of nice lengths of timber along the way.

It seems incredibly quiet now on the beach with the cable ashore.

The Nordica is away out past the island at St.Ives gradually diminishing towards the horizon, tipping out cable as she goes, enroute for the wave hub marina. The hub itself is on board, a bright yellow thing.

Two fluorescent coated men were also strolling on the beach, happy to pass the time of day in the soaking wetness.

"We're on weather watch," they said laughing as they told me they were working for the firm the client had contracted to oversee the whole project.

"Some of us are on the boat." "Wow," I said enviously imagining the privilege.

"Oh no," they warned. "That's no fun, they've been stuck on board there since Hartlepool, 6 weeks with nowhere to stretch your legs and tiny cabins."

Put like that I could imagine the hardship.

"But what a trip, coming around the north!" I said, unable to resist my imagination.

"That's not a trip" One replied, as the water dripped off the end of his nose, "That's a nightmare, believe me I was in the navy and it's horrible by boat, but anyway this boat came around the other way."

Ah the dangers of hearsay, misinformation and small town chinese whispers.

What I did discover however was that tomorrow or maybe the next day which seem now to count as the same thing except with a different number of sunrises and sets, the cable laying team are burying the cable with a state of the art air/water drilling gadget that functions underwater using pressure to wash the sand back over the cable. The whole procedure should only take one hour to bury the 1.8km length to a depth of 2.5m. High speed.

We shall see.

I asked about the previous project managers and if they had got brutally axed for failing to deliver the cable on schedule. The men bluffed reassuringly with only the merest flicker of doubt passing over their pleasant wet faces, shining beneath the hard hats.

How much to tell this strange woman?

They explained that the old team were simply being deployed on other jobs to make space for another team to have a crack at the whip. Of course there had been all sorts of technical challenges such as the depth of the tide and the shifting sands.

I told them what I know of the beach and the sand and the Harbour and they threw some stones for the dogs who had by this time taken up their impatient barking before deciding it was time for their coffee.

We parted on see you tomorrow terms for the burying of the cable: the final incarceration.

※

Walking towards the group of men changing their clothes in the car park it was clear from a distance that the atmosphere was already friendly.

"Are you the cable layers and trench diggers?" I asked as if I owned the place.

"Yes," they said. "We are. Tomorrow will probably be the day."

"Is it true you do it underwater?" I asked.

The Wave Hub Chronicles

"Yes," they said. "We do. A high pressure water jet cuts through the sand, the cable goes in and the sand closes over it."

"So tell me," I asked. "What do you wear?"

Eyebrows raised.

"Well we're in a boat with a motor and the jet goes along beside."

"Oh" I said a little disappointed, "You're not actually in the water?"

So they explained how it was and I, with a surge of queenly feeling, said.

"All the best, you may get on with it."

And they played along and with a wave of the regal hand we laughed and that was that, merry and sweet.

A little while later we saw them on the beach and they grinned and gave the thumbs up as they stood around the big digger with the caterpillar tracks waiting for tomorrow.

The Nordica still hovers by the Island at St.Ives.

I got back from a day out at a rocky cove on the South Coast in glorious blue gold sunshine to remember that I'd totally forgotten about the sinking of the cable and thought "darn, I've missed it."

This morning, however before going away, I nipped over to the dunes to see what was on and there, in the water, were a few small boats, a floating pontoon and a spray of water spurting up from the sea.

Beside me at my vantage point was a man with a video camera filming the scene. He has been employed by the RDA to document the events for the wave hub project archives.

The Wave Hub Chronicles

The final laying of the cable was scheduled for high tide this evening.

This could fit in tidily with my high tide swim.

Later, returning from town on my bike I met the weather watchers this time sweltering in the sunshine, their faces shiny with sweat beneath their plastic hard hats.

They informed me that the final cable laying would in fact be tomorrow to ensure the resettlement of the sand over the cable to sufficiently withstand the surge of the tides.

The Nordica was sitting pretty by St. Ives, waiting for the word to go after the cable was buried for good and all.

I suggested that slices of the cable might be worth something on ebay and although they laughed at first their attitude towards me changed, becoming a little cooler and more suspicious.

Funny that.

※

It is bright. It is a blustery bank holiday weekend. Families lie about, kite surfers fly across a blue horizon fringed with frilly surf. The Nordica sits on the sharp horizontal line waiting to continue her sea bound cable laying mission. She will travel at the maximum speed of 200 metres per hour. A couple of smaller tugs wait around nearby to lend support. On the beach the mechanical diggers are burying the remains of the cable by the coffer.

A friendly cable layer from Yorkshire chats to me about the process as water and sand slurry gurgle over the cable and it disappears into the trench. We gently enter the subject of sustainability and respect for nature , most particularly the Sea. The conversation softens into a trustworthy companionship, agreeing wholeheartedly on the futility of built in obsolescence.

I feel a motherly tenderness towards this young man who has no reason to be at odds with the world and yet finds himself in a culture that attempts to dupe the spirit into believing in the capitalist nightmare at the expense of common sense and generosity.

One of the things I think I object to most about the capitalism we live with, notwithstanding the historical theft it is founded upon or the people annihilated because they stood in its way, is the insistence on perpetual growth. With reference to the five element system the season that is associated with growth is the spring, when everything comes burgeoning forth after the deep rest of winter. It is a beautiful hopeful time but if it is not allowed to bear fruit, mature and then decline and die with equal respect for each phase of its existence, then we start to witness the sort of world that is at present our cultural reality.

Other seasons' qualities become downgraded and discredited; parenting is potentially very isolating, youth have little wise counsel because old age becomes burdensome and a sign of decrepitude, (best get a face lift or botox injection in case a wrinkle might give away the fact you've been around the block.) and death becomes positively taboo.

Perpetual growth is also a nightmare when it surfaces as cancer with cellular frenzy unable to regulate itself.

The young man's colleague approaches suggesting lunch and I'm surprised to discover that neither of them have yet sampled a Cornish pastie.

The Wave Hub Chronicles

September

The digging continued through to the end of the holiday and at last the cable was totally buried right up to the coffer.

The security men have been laid off from the job, no more to emerge from the metal container to smoke yet another cigarette in the seaside air.

It's curiously quiet everywhere. No fluorescent jackets are flashing through the marram grass, no sense of anything impending or tide depending. No Nordica sitting waiting.

Back on the beach the last of the families soak up the sun. School is back on and the morning and evening air is crisp with autumn promise.

Surf's up good and proper with a warm off shore wind and the potential of a barrel if you were really quite short.

All that remains is to connect up the cables inside the coffer and keep laying the cable away out to sea.

It's all being done out of view from our chalet's windows or my beachside observations.

The sky is full of feathered clouds, which fly across the blue as a large flock of birds wing away beneath them.

The birds are gathering, the workmen departing, the next phase is being heralded in for when the mighty ocean clocks up her wattage and the hub channels it through.

I wonder if the quality of power will be perceived at the receiving end?

I fancy I can tell the difference between water heated by my Raeburn and that by the electric element.

Perhaps thousands of households may find their minds wandering to the dark mysterious ocean as they draw the curtains and switch on the light.

※

Long low wet clouds splay over the September chill.

The sound of grinding metal penetrates from over the back.

One fluorescent man strolls near the coffer.

One carriage on the St.Ives train remains from the holidays four.

The Family holidays are over for another year. Its now the season for the newly weds and nearly deads.

Harvest time arrives with a feeling of completion, a time for storing up and closing down the extrovert summer.

※

Swimming in a huge harvest high tide sea, the coffer is surrounded like a rusty island.

All of a sudden an orange man appears at the top of the wall and climbs out from the inside, followed by another.

I shout "Ahoy there!" in the rain, but they don't hear and climb down a ladder on the outside into at least a couple of feet of swelling water.

It must be knocking off time. They must have been in there connecting the cables ...

The Wave Hub Chronicles

※

This beautiful morning follows a dazzling starry night and rainy dawn.

The beach is swept clean by the tides.

※

Over these next few days the tides reach right up to the dunes and retreat miles out to the most extreme edges of their spectrum. The dunes are eroding before our eyes. A couple, who have holidayed here for the past 40 years were all ready to blame the Wave Hub.

As I defended it, surprising even myself, I couldn't help wondering about the effects of the drilling and the coffer acting as a groyne in the water.

"However" I cried vehemently," You can't go relentlessly digging and removing sand from the beach without consequences."

I am angry.

Pissed off and feeling powerless.

Please just leave it all alone.

Lets stop treating our world as an expendable commodity.

※

On this sunny warm day the clouds are casting shadows on the huge beach.

They are back digging away at the base of the coffer.

I recognize the dumper truck driver with the wet boots and jump a dug out gully, missing the edge and losing my shoe in some soft sand and plunging my apple into it as I fall over.

"What are you doing now?" I ask of him, recovering my balance if not my composure.

He turns and tells me that they are trying to pull a bit more cable through as they didn't quite leave enough to meet up and make the connection.

My mind does a series of mini jigs as I stifle a snigger at how ludicrous this sounds imagining a few of the engineers tugging on the ten miles of it to give that extra few inches.

I walk barefoot to the low tide mark, which is as low as it will get all year so they say and arrive at a spit of sand, which seems literally miles away. The river is broad and wild out here, undulating against the seas currents, humping and falling in strong crescendos as sand at the edges cracks and falls away into the water, unable to stand with itself, a miniature reenactment of the cliffs erosion.

Following the river back towards the shore I climb the repeated swirls and humps left by the water's dance. Little pools of water contain the cracked limbs of crabs. They remind me of filter tipped fag butts in a soggy ashtray. I feel suddenly very weary and sick of the mess we're making.

☼

After our last night's fire we go to jump in the high tide sea this morning.

Although the tide's already turned a while it's still well up and deep for swimming. Bobbing in the fresh blue waves, the morning sun blasts shadows from the dunes across the water and the wind buffets all the body parts that are outside the water. Those submerged feel the pull of the currents.

The Nordica comes into view at the horizon back from her cable laying adventures. Blue and white she pushes a line just below the horizon, to come to rest in the bay.

Later in the slanting evening brilliance, as the tide rises once again she appears to have a tall silver chimney sticking up at her rear behind the bridge. I don't think that was there before. There's no one around to ask and I can only imagine that she might be shimmying the cable along to get rid of the shortfall at the connecting point, like pulling up a wet suit over wet legs and feet so that you can get the top half on.

The sun is now setting down behind Carbis Bay, getting closer to its quarter mark between the solstices.

※

Actually the Nordica has just been in Portland for repairs, one of its thrusters got broken. Its out there now digging trenches for the remaining bits of cable, the tall silver chimney thing is part of the trenching machinery. Another boat is out there laying rocks on the hub and other strips of cable, but every few days they have to go and fetch more rocks from France because our native rocks are somehow wrong.

The Hub is in position and a new one has been ordered by Spain.

All this information came from the cable layers, so and so umbilical services who make cable and fibre optics. They've been told down the pub that plenty of people are keen to come and dig up the cable.

"You're the boss," they told me strangely, "Go and write it down."

"That's the boss." I said pointing to the ocean that I had just stepped out of.

I stand in my toweling robe, fresh from the dance in the waves where I span about looking up to the sky, supported by the amazing buoyancy of the water with the land cradling us all around.

The cable layers were the two men I'd seen climbing out of the coffer and getting wet a few days ago.

Connection should be made later this week.

※

The reptile fence has gone.

The blackberries are fruiting above a dewy ground. The air is cold and warm in wafty differences. There is the sound of drilling coming from inside the rusty coffer and a lot of transit vans and men in orange congregating in the car park. Their logo proclaims "Power man", with a fist clenching a red line.

※

These days the power men park regularly.

We can't help but wonder if there are any power women among them.

They're on the beach now making the connection.

※

The power men were on the beach early again this morning.

The kelp is in and strewn all along the high water mark. The dogs were jumping for the stalks.

They met another dog belonging to an older man with the workforce who stands in front of the blue metal container which functions as both a tea hut and security booth. Two younger men in orange red overalls were standing out next to the blue. Glazy eyed, London accented they were cool, laconic, neither friendly nor unfriendly.

Involved in the business of making connections they are down in the rusty coffer, in the damp sand and concrete introducing wire and fibre optics to each other with invisible mends. There was another lad from Scotland and oh don't forget the Welsh one they told me. The older man with the dog was an urbanite whose inscrutable opinion on sitting on a beach was beyond me.

As I walked away past the litter of kelp I didn't pass the time of day with these guys; they seemed to be operating in a bubble, nature of the job perhaps.

All that testosterone in close proximity with all that electricity. Who knows maybe it creates some particular electro magnetic chemical field?

※

The red hot pokers are blooming proudly.

Great piles of sand next to the coffer afford access for the diggers to reach in and down with their mechanical arms.

Now the coffer sits quiet and abandoned in the still grey autumn air.

The Wave Hub Chronicles

October

Misty and murky the damp coffer's metal darkens rusty, the colour of a conker. They've taken away the generators from their perches bracketed onto the sides.

※

This blustery morning's high tide dares me to enter the chilly water as the wind buffets about. There are some men standing around the coffer but I decide to benignly disregard their presence and do my tai chi in full view in the middle of the beach before jumping in the waves.

The beach dog ban has expired now we're past September so my tai chi practice is reestablished in view of the water. During the summer I practice over the back in a grassy clearing among hawthorn, bramble and honeysuckle where the bunnies and dogs play life threatening hide and seek, which ends in dinner or not …

Now back with the Sea and the Sky I feel the huge space inside me.

As I swim bobbing in the swirling brine one of the men waves. I'm a bit shortsighted so can't be sure but I reckon he's the friendly diver who put the coffer up.

Showering after my dip I realise I've become one of those plucky older women who go in the sea in all weathers. In my mind they belong somehow to the older generation: Identity is a strange thing if you don't keep up with the changes.

※

In glorious late afternoon sunshine we walked past the coffer. The rusted iron gleamed so beautifully in all its burnished splendour.

Earlier some men had sliced off half the height of the back wall with some kind of heat. Bubbles of molten iron dried silvery at the cutting edge.

We clambered onto a mound of sand to peer down into the depths. It was not in fact too far down inside the metal girders that prop the walls of sand at bay, bracing against the force to crumple inwards. There was about 2 foot of water drowning the oblong connection box, coated in resin.

※

Next morning that spatters with rain, I discover the big digger has left the same massive braces by the side of the track up to our chalets.

Later on, the screeching sound of giant metal teeth being pulled reverberated through the dunes as the digger with claw extracted the girders and rendered them horizontal on the beach.

Walking towards this activity from a few hundred metres' distance, I could see a man hitting the metal girder with a sledge hammer, his actions happening a full half second before the sound reached my ears like a badly dubbed movie out of synch, causing me to question time distance sound and light waves in one of those wonderful nano moments that we habitually take for granted.

The Wave Hub Chronicles

Witness again the extraordinary way in which the mind can travel over continents, through history and walls in the blink of an eye.

Over the back they have been raking gravel around the new transformer as neatly as around a well tended grave.

※

The beach is now back to its wild self: no flags, no windbreaks, no hoards of people, just the odd one or two, heads down against the wind or backs to it, moving quickly, blown along like sea foam.

One big tide and it'll be as if the workforce were never here. All the tracks washed by the waters as delicately as a tear brushed from a cheek.

※

Its been almost a year give or take a few weeks since it all began and the rain has darkened all the leaves, sand and wood, staining them with autumn and the encroaching dark time.

※

Cycling back up the track in a rare moment of post rain sunshine, I hear a tinkling of metal as a digger rattles its way over the beach.

Peering through a gap in the bushes I spy the digger, a chain dangling from its arm, attached to a seesawing iron girder. On the beach working late a couple of guys wait by the pile of girders still to be scooped up and delivered back round to base.

As I walked out to take a dip I asked, "Is this the end of it then?"

"Yes" They said, "It will be."

They didn't think the sand would move to reveal the cable in its beachy grave.

"It's in too deep."

By Christmas or January the current would be flowing and switched on.

"Mind you" said one "This wasn't supposed to take this long, so you never know."

He threw a stick for my barking dog, the digger arrived back for another load and I headed for the waves as the golden sun appeared, descending beneath a dark grey rain cloud.

***

The two men sat in the autumn sunshine by the remains of the project: skips, porta cabins, piles of unused materials: Bricks, felting and wood lay around. They were just finishing up the tidying up and agreed that I could take some of the wood that had protected the massive coils of cable, for our fire or even bits of building. They had really enjoyed the job and were in no hurry to be gone in fact they were maybe even a little sad, especially about leaving the beach.

"We've been around since March, that's seven months. Yeah, it's been good," one said wistfully, "they should be switching on in the new year."

I didn't want to spoil his futuristic nostalgia, but wondered which month of the New Year he might mean, even if it were actually in this new year and not another, some few hence.

When I returned to pick up the wood one of them was riding the roller, rollering the roadway up towards the substation building.

The other was bent over an angle grinder slicing slabs off the cable into salami style pieces. I laughed, thinking they'd taken up the idea of flogging bits of the cable on ebay, but he said they were for display in the office and they did indeed look pretty: three shiny copper circle and two fibre optic clusters all in circular cross section on the piece that was from the seaward cable.

The spare length of cable from the sea was only a few feet, after all those miles they'd cut it exceedingly fine.

The Sea at high tide was wild and gnarly, full of force, incredible swell and challenging surf. Bobbing about in the waves I felt thrilled and slightly nervous in the power of it. The tide is hugely high with the new moon coming and finally all remnants of digger tracks are washed clear as the sea slaps all the grains of sand into alignment erasing memories with one swipe.

☼

There's a drummer who has been coming to drum to the river as the egret stands secretly watching a flock of Canadian geese gathering.

So much talk of the dune's erosion as sea buckthorn floats in the swollen tide. Bitten off by hungry waves it is spat out and spewed on the beach in tangled clumps.

The Wave Hub Chronicles

November

The extremely high tides continue to scour away at the dunes, dumping sea buckthorn into the waves in huge thorny clumps, lethal flotsam to the unsuspecting swimmer. Huge wadges are lying on the beach entwined with a million plastic bottles and other garbage.

We made a huge blaze on the beach picking pieces directly from the sea to feed the fire. Once the flames were hot enough it all got gobbled up.

For a few days now the junction box has been visible, sticking up from out of its sandy bed like an unexploded 2nd world war bomb.

Nearby a sawn off remnant of the metal coffer juts out, a stubborn tooth refusing extraction, defiantly ancient and rusty it pokes through.

I went to the wave hub office to let them know if they didn't already so they sent out their marine engineer, who is just down from the Isle of Skye. He is used to the Cruel Mistress as he calls her, understanding the power and respect due to the Sea. Working on the oil rigs in the North Sea in the 1970's and 80's, they earned stupid money in wild conditions.

He tells me of the seven currents just off the coast here which make it particularly attractive for wave harnessing and of the high octane history of Hayle from WW2 which left the legacy of a huge electricity cable running from Manchester to Indian Queens and from there a sub cable to Hayle.

He tells me they will reexcavate to sink the connection box a further 2 metres.

A young man in a red sports car has been along with a shovel. He dug away all afternoon.

We had hoped that maybe the Wave Hub office could use their influence to persuade the Harbour company to put back some of the stolen sand.

Last year they dumped a whole load up in the dunes, it was completely out of the system and so irrelevant to the closed cell system that operates in the bay.

All day long the wind has been blowing, bringing squalls of rain, interludes of blue sky , drenchings, door slammings and mild irritation.

The beach has been sluiced smooth by the high tide, the young man's work dismissed with a wave, hours of shoveling rendered obsolete, the connection box poking back out, like a cheeky tongue.

Windsurfers are flying across the horizon line, typewriter carriages returning to a new paragraph, rainbows twinkling on, then off into bleary drizzle. The buffeting continues and feels endless.

I sneak into a lecture at Truro College about the Wave hub, given by the head of Marine energy at the RDA.

It contains lots of interesting business and technical information and a glimpse into the world of energy provision that turns my mind numb but makes me feel like I'm doing my research. At one point he hands around

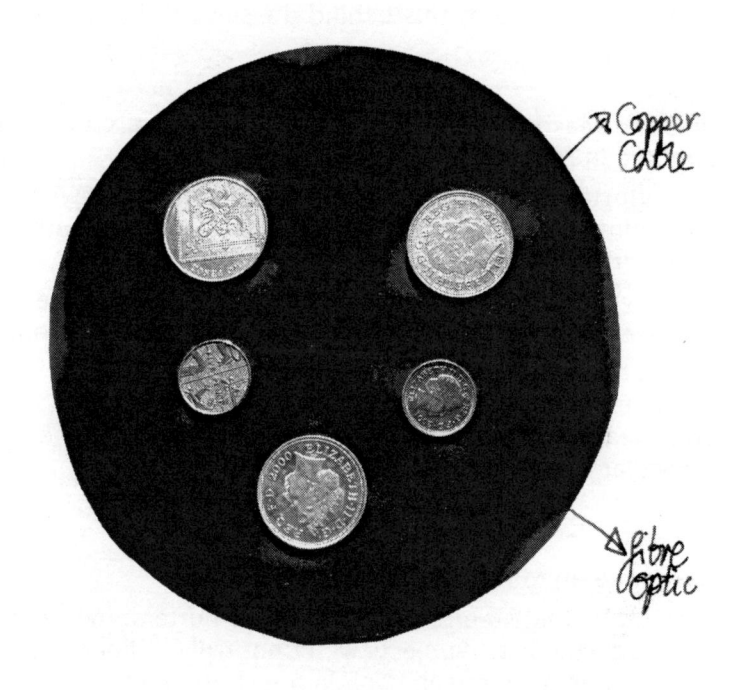

Artists impression of a cross section of the seaward going cable.

the salami slices of cable that I saw being cut and I feel stupidly privileged by this behind the scenes knowledge.

It is the season of rainbows in the bay as squalls and light combine in refracting mystery.

The junction box and cable lie bare and exposed awaiting their reinterrment by experts from Exeter and Plymouth Universities.

A tall yellow buoy has stood on the quayside for some days waiting to be taken out to mark the wave hub's spot.

When the actual buoys will be ready is any one's guess. Funding for research apparently mostly comes from wealthy individuals with excess income and some particular interest.

I want again to mention the spiritual.

The realm that so often is reserved for nutters, hysterics or fundamentalists. Subjects we're not truly at liberty to discuss unless we're someone in the church.

The ownership of the Divine has already been mortgaged out to someone who will sell a bit of it on to you.

Real raw access, direct experience has been deemed too dangerous, and has become the domain of the magician. Although the rise in interest in Paganism indicates our thirst for connection.

Our personal glimmers and understandings of the extraordinary mystery we inhabit are so often marginalized or ridiculed. The Heart Path is not just the stereotypical hippy choice of dropping out from responsibility, rather a living choice to directly relate with the world, listening carefully to the lessons of our lives and taking full responsibility for the learning.

How else can we possibly sustain ourselves?

The Wave Hub Chronicles

Bowing down to the awesome powers stretches and opens our spinal chords, the nerve centres of our own divinity. The most important changes we need to make are in attitude.

※

There are people arriving from the Universities to dig into rock and take core samples to see if further excavation is viable.

Meanwhile, the sand erodes from further back in the dune.

The earliest snow in 17 years has fallen over the West Country, everything is in snowy deep freeze.

I awoke to a flash of lightning, a clap of thunder then blizzard snow, all west of Truro.

Beyond Truro, east was clear blue sky and green.

Walking along the beach and passing through a cave a lump of rock falls off just behind me, thudding dramatically into the sand, just glad it wasn't my head.

The Wave Hub Chronicles

December

The frozen week warmed up to rain as the wind changed to a south westerly. Drab dank wetness replaced the frosty brightness but now its clear and blue and gold again. Along by the estuary a heron and egret stand off for territory as ducks paddle by. The Heron wins the area.

Three men are standing by the gate. They've been there since 7.30am waiting for someone to unlock the gate and show them the job. They know nothing of the Wave Hub, simply that something needs digging in, there's trouble with a cable. One of them has piercing blue eyes. They are all proper Cornish.

"Seems like they've got money to waste" He says archly as I leave them waiting.

There's no sound of any work later on, we'll see tomorrow ...

There is another incredible deep freeze. Jack Frost is sparkling everywhere bejeweling the drabbest of articles, even piles of dog pooh. Amazing crystal clear light casts brilliance and shadows sharply.

Forests of kelp lie felled in the rolling sand's hills.

I move through my Tai chi by the little cove that is now banked up smoothly.

Three large mechanical diggers trundle over the horizon and start to dig around the connection box. They dig as I write. Is the pen mightier than the scoop? After all the contractors have come and gone it is now down to the three local men to bury the lot another 2 metres down. Two and a half metres down is rock. Apparently in the last six weeks the beach has lost 3 metres of sand, it's all over at St. Ives or just around the corner so the officials tell me, except of course the lorry loads that went off up the A30.

January

About 20 foot of black pipe the dune side of the connection box has popped up through the increasingly gravelly sand.

A couple of weeks later sand has come back in to lay it to rest.

And so it goes on.

Because the works have finished on the beach and over the back so are these personal chronicles.

The Wave Hub will hopefully be plugged into in the not too distant future and the Ocean's blessing will be reaching your plug sockets soon.

The Wave Hub Chronicles

Postscript

It is now nearing Spring 2011.

There's been no further news of the power buoys and the Hub.

Japan has just experienced the worst earthquake and tsunami in living memory.

Britain is in the grip of severe job cuts, escalating fuel and food prices and a palpable sense of uncertainty.

The Local newspaper reports so many stories of crimes fuelled by alcohol or drugs but rooted in emotional misery.

National news features politicians who seem so fork tongued its difficult to trust them.

The poles of the Earth's axis are shifting, there are sunspots flaring.

Discrepancies in equality around the globe manifest an appalling lack of balance.

And there are also many wise and wonderful beings working towards the healing and balance of our people on the planet.

I sometimes find it difficult to know what I can do to make a difference. I feel it always returns to the personal.

"If there is peace in the heart,
There will be beauty in the person,
If there is beauty in the person
There will be harmony in the home,
If there is harmony in the home
There will be order in the nation,
If there is order in the nation
There will be peace in the world."

Holistic medicine helps validate us as spiritual beings as well as physical bodies with minds and emotions and is well worth experiencing to help in the general healing process that seems so necessary for harmony and balance.

My favourites are Plant Spirit Medicine (plantspiritmedicine.org), Homeopathy, Acupuncture, Herbalism and Cranial Osteopathy.

Sitting around a Sacred Fire helps bring us into our Hearts, a sense of community and burns away the unnecessary.

For info on a fire near you: sacredfirecommunity.org.

The practise of an inner relaxation technique such as Tai Chi, Yoga or meditation can be very helpful.

Good Luck in finding your own way. Thank you for reading this account of a year on the beach.

Stop Press

Just the other day, moments before this book is sent off to the printers, I went into the local hardware shop where they are selling calendars with photos from the Hayle Community Archive. The calendars include a number of photos from the beginning of the 20th century. There is the old power station with huge smoking chimneys set amidst massive piles of coal, ships and conveyor belts. The scene describing the industrial innovations of the time. The natural backdrop is an absolute irrelevance as electricity is produced locally for the first time. In conversation with the shopkeeper I learnt that the dunes on which we live were massively augmented, if not totally created, by all the sand and rock and who knows what else dug out so that the foundations of the power station could be laid. For me this introduced a whole other slant on the story I have told so far: Not only was the UBO so likely to be something buried by our Victorian forebears but also there is an incredible sense of continuity from one century to another. There will always be repercussions from our actions. It also had me feeling that I am only just starting to know this place and all the twists and turns that create it – both natural and human made. My observations in this book could so easily have been called simply "A year on the Beach."

About the Author

Lucy Wells is a practicing Plant Spirit Medicine healer, a long time Taichi practitioner, a sometime community artist, maker and performer.
She is an initiated Firekeeper and currently lives by the sea in Cornwall.

Acknowledgements and gratitude

Thank you to Michael Locke whose Love and support have been so much more than good enough.

Thank you to Mani Layward Wells, my main critic who keeps me on my toes.

Thank you to Joan Wells, the best read woman I know, for coming up with the quote after reading the first draft: "Let's face it, it's not Shakespeare but I couldn't put it down."

Thank you to Janet Mcewan for taking the cover photograph with her inimitable style.

Thank you to Jonathan How for helping me with all the challenges of self publishing, just like a proper publisher.

Thank you to my teacher of Plant Spirit Medicine, Eliot Cowan for all that he has introduced me to.

Thank you to all my Tai chi and Qigong teachers along the way for their dedication and generosity.

Thank you to all the people who were working on the Wave Hub who bothered to chat and share their knowledge.

※

Thank you to our Grandmother Ocean and Grandfather Fire and to all our relations past, present and future.

Lightning Source UK Ltd.
Milton Keynes UK
UKOW021900031011

179705UK00001B/5/P

9 780956 936790